Serverless Machine Learning with Amazon Redshift ML

Create, train, and deploy machine learning models using
familiar SQL commands

Debu Panda

Phil Bates

Bhanu Pittampally

Sumeet Joshi

BIRMINGHAM—MUMBAI

Serverless Machine Learning with Amazon Redshift ML

Group Product Manager: Ali Abidi

Publishing Product Manager: Ali Abidi

Book Project Manager: Farheen Fathima

Senior Editor: Tazeen Shaikh

Technical Editor: Rahul Limbachiya

Copy Editor: Safis Editing

Proofreader: Safis Editing

Indexer: Tejal Daruwale Soni

Production Designer: Prashant Ghare

DevRel Marketing Coordinator: Vinishka Kalra

First published: September 2023

Production reference: <1131015>

Published by Packt Publishing Ltd.

Grosvenor House

11 St Paul's Square

Birmingham

B3 1RB, UK.

ISBN 978-1-80461-928-5

www.packt.com

To my wife, Renuka, and my kids, Nistha and Nisheet, who make my life exciting every day!

– Debu Panda

To my wife, Patty, who encourages me daily, and all those who supported me in this endeavor.

– Phil Bates

To my wonderful family – my wife, Kavitha, and my daughters, Vibha and Medha.

– Bhanu Pittampally

To my wife, Leena, my daughter, Ananya, my parents, and my brother, Neeraj, for their support and encouragement for all these years – especially my wife, who always supports me in all walks of my life.

– Sumeet Joshi

Foreword

In today's fast-paced technological landscape, the convergence of serverless computing and machine learning has transformed the way we approach data analytics and decision-making. As the demand for scalable solutions continues to rise, the need for accessible tools that bridge the gap between complex algorithms and user-friendly interfaces has become paramount.

In *Serverless Machine Learning with Amazon Redshift ML*, the authors embark on a journey that empowers users, regardless of prior machine learning experience, to harness the power of data-driven insights. By leveraging simple SQL commands, this book will walk you through how to solve complex problems using different machine learning algorithms with ease. Gone are the days of steep learning curves and complex coding – this book paves the way for a new era and describes how Amazon Redshift ML democratized machine learning.

Through a balanced approach of theory and hands-on exercises, this book guides you through the fundamentals of machine learning concepts while showcasing how Amazon Redshift ML serves as the cornerstone of this revolutionary approach. The authors walk you through the benefits of serverless computing, demonstrating how it not only enhances the process of training your machine learning models but also streamlines the entire process.

Whether you're a seasoned machine learning professional or are just starting on your machine learning journey now, this book will provide the roadmap you need. The authors are authoritative sources on the topics who defined Redshift ML and work with customers to make them successful with a variety of use cases, such as product recommendation, churn prediction, revenue forecasting, and many more. Enjoy your journey to machine learning and allow the book to unlock the potential of machine learning in your Amazon Redshift data warehouse one simple SQL command at a time.

Colin Mahony

GM, Amazon Redshift, AWS

Contributors

About the authors

Debu Panda, a senior manager in product management at AWS, is an industry leader in analytics, application platforms, and database technologies, and he has more than 25 years of experience in the IT world. Debu has published numerous articles on analytics, enterprise Java, and databases, and he has presented at multiple conferences, such as re:Invent, Oracle Open World, and Java One. He is the lead author of *EJB 3 in Action* (Manning Publications, 2007 and 2014) and *Middleware Management* (Packt Publishing, 2009).

I want to thank the people who supported me, especially my wife, Renuka, my kids, Nistha and Nisheet, and my parents.

Phil Bates is a senior analytics specialist solutions architect at AWS. He has more than 25 years of experience implementing large-scale data warehouse solutions. He is passionate about helping customers through their cloud journey and leveraging the power of machine learning within their data warehouse. Phil has written several blogs on Amazon Redshift ML and presented at re:Invent and AWS summits. He enjoys golf and hiking and lives with his wife in Roswell, Georgia.

Bhanu Pittampally is a seasoned professional with over 15 years of invaluable experience in the realm of data and analytics. With an extensive background in data lakes, data warehouses, and cloud technologies, Bhanu has consistently demonstrated a deep understanding of the ever-evolving landscape of data management.

Armed with a wealth of knowledge and two advanced degrees – a master of science and an MBA – Bhanu combines academic rigor with practical insight to provide his clients with comprehensive solutions to their intricate worlds of data.

Sumeet Joshi is a solutions architect/data scientist based out of New York. He specializes in building large-scale data warehousing and business intelligence solutions. He has over 19 years of experience in the data warehousing and analytical space.

About the reviewers

Anand Prakash is a senior data scientist at Amazon, based in Seattle, USA. He has a strong passion for technology solutions, particularly in ML, MLOps, and big data. Anand is always eager to learn and grow professionally, constantly seeking new knowledge and opportunities. He shares his knowledge on various tech topics through his writing at `https://anand-prakash.medium.com/`. He holds a bachelor's degree in electronics and communication from the **Northeastern Regional Institute of Science and Technology (NERIST)**, Arunachal Pradesh, India.

Anusha Challa's journey as a data warehousing specialist has allowed her to work with multiple large-scale customers of Amazon Redshift and AWS analytics services. Along this path, she's had the privilege of collaborating with brilliant minds, while the steadfast support of her cherished family and friends has been her constant driving force. Keen to fuel her spirit of continuous learning, she finds inspiration in working with data – a realm that perpetually evolves and expands – making each step in data warehousing and ML an exciting journey of exploration and growth.

Guided by insights gained from her master's degree in ML, she is happy to contribute a review of this book, which presents diverse methods integrating data warehousing and ML, thus broadening the horizons of what data warehouses can encompass.

Table of Contents

Part 1: Redshift Overview: Getting Started with Redshift Serverless and an Introduction to Machine Learning

1

Introduction to Amazon Redshift Serverless   3

2

Data Loading and Analytics on Redshift Serverless   25

6

Building Classification Models   105

7

Building Regression Models   131

8

Building Unsupervised Models with K-Means Clustering   157

Part 3: Deploying Models with Redshift ML

9

Deep Learning with Redshift ML   173

10

Creating a Custom ML Model with XGBoost   187

11

Bringing Your Own Models for Database Inference   205

12

Time-Series Forecasting in Your Data Warehouse   221

13

Operationalizing and Optimizing Amazon Redshift ML Models   237

Preface

Amazon Redshift Serverless enables organizations to run petabyte-scale cloud data warehouses quickly and in a cost-effective way, enabling data science professionals to efficiently deploy cloud data warehouses and leverage easy-to-use tools to train models and run predictions. This practical guide will help developers and data professionals working with Amazon Redshift data warehouses to put their SQL knowledge to work for training and deploying machine learning models.

The book begins by helping you to explore the inner workings of Redshift Serverless as well as the foundations of data analytics and types of data machine learning. With the help of step-by-step explanations of essential concepts and practical examples, you'll then learn to build your own classification and regression models. As you advance, you'll find out how to deploy various types of machine learning projects using familiar SQL code, before delving into Redshift ML. In the concluding chapters, you'll discover best practices for implementing serverless architecture with Redshift.

By the end of this book, you'll be able to configure and deploy Amazon Redshift Serverless, train and deploy machine learning models using Amazon Redshift ML, and run inference queries at scale.

Who this book is for

Data scientists and machine learning developers working with Amazon Redshift who want to explore its machine-learning capabilities will find this definitive guide helpful. A basic understanding of

machine learning techniques and working knowledge of Amazon Redshift is needed to make the most of this book.

What this book covers

Chapter 1, Introduction to Amazon Redshift Serverless, presents an overview of Amazon Redshift and Redshift Serverless, walking you through how to get started in just a few minutes and connect using Redshift Query Editor v2. You will create a sample database and run queries using the Notebook feature.

Chapter 2, Data Loading and Analytics on Redshift Serverless, helps you learn different mechanisms to efficiently load data into Redshift Serverless.

Chapter 3, Applying Machine Learning in Your Data Warehouse, introduces machine learning and common use cases to apply to your data warehouse.

Chapter 4, Leveraging Amazon Redshift Machine Learning, builds on *Chapter 3*. Here, we dive into Amazon Redshift ML, learning how it works and how to leverage it to solve use cases.

Chapter 5, Building Your First Machine Learning Model, sees you get hands-on with Amazon Redshift ML and build your first model using simple CREATE MODEL syntax.

Chapter 6, Building Classification Models, covers classification problems and the algorithms you can use in Amazon Redshift ML to solve these problems and learn how to create a model with user guidance.

Chapter 7, Building Regression Models, helps you identify whether a problem involves regression and explores the different methods available in Amazon Redshift ML for training and building regression models.

Chapter 8, Building Unsupervised Models with K-Means Clustering, shows you how to build machine learning models with unlabeled data and make predictions at the observation level using K-means clustering.

Chapter 9, Deep Learning with Redshift ML, covers the use of deep learning in Amazon Redshift ML using the MLP model type for data that is not linearly separable.

Chapter 10, Creating Custom ML Model with XGBoost, shows you how to use the Auto Off option of Amazon Redshift ML to prepare data in order to build a custom model.

Chapter 11, Bring Your Own Models for In-Database Inference, goes beyond Redshift ML models. Up to this point in the book, we will have run inference queries only on models built directly in Amazon Redshift ML. This chapter shows how you can leverage models built outside of Amazon Redshift ML and execute inference queries inside Amazon Redshift ML.

Chapter 12, Time-Series Forecasting in Your Data Warehouse, dives into forecasting and time-series data using the integration of Amazon Forecast with Amazon Redshift ML.

Chapter 13, Operationalizing and Optimizing Amazon Redshift ML Models, concludes the book by showing techniques to refresh your model, create versions of your models, and optimize your Amazon Redshift ML models.

To get the most out of this book

You will need access to an AWS account to perform code examples in this book. You will need either to have administrator access or to work with an administrator to create a Redshift Serverless data warehouse and the IAM user, roles, and policies used in this book.

Software/hardware covered in the book	Operating system requirements
The AWS CLI (optional)	Windows, macOS, or Linux

If you are using the digital version of this book, we advise you to type the code yourself or access the code from the book's GitHub repository (a link is available in the next section). Doing so will help you avoid any potential errors related to the copying and pasting of code.

Download the example code files

You can download the example code files for this book from GitHub at `https://github.com/PacktPublishing/Serverless-Machine-Learning-with-Amazon-Redshift`. If there's an update to the code, it will be updated in the GitHub repository.

We also have other code bundles from our rich catalog of books and videos available at `https://github.com/PacktPublishing/`. Check them out!

Conventions used

There are a number of text conventions used throughout this book.

`Code in text`: Indicates code words in text, database table names, folder names, filenames, file extensions, pathnames, dummy URLs, user input, and Twitter handles. Here is an example: "Mount the downloaded `WebStorm-10*.dmg` disk image file as another disk in your system."

A block of code is set as follows:

```
cnt = client.execute_statement(Database='dev',
    Sql='Select count(1) from chapter2.orders;',
    WorkgroupName=REDSHIFT_WORKGROUP)
query_id = cnt["Id"]
```

When we wish to draw your attention to a particular part of a code block, the relevant lines or items are set in bold:

```
SHOW MODEL chapter5_buildfirstmodel.customer_churn_model;
```

Any command-line input or output is written as follows:

```
$ pip install pandas
```

Bold: Indicates a new term, an important word, or words that you see onscreen. For instance, words in menus or dialog boxes appear in bold. Here is an example: "Select **System info** from the **Administration** panel."

> **Tips or important notes**
> Appear like this.

Get in touch

Feedback from our readers is always welcome.

General feedback: If you have questions about any aspect of this book, email us at customercare@ packtpub.com and mention the book title in the subject of your message.

Errata: Although we have taken every care to ensure the accuracy of our content, mistakes do happen. If you have found a mistake in this book, we would be grateful if you would report this to us. Please visit www.packtpub.com/support/errata and fill in the form.

Piracy: If you come across any illegal copies of our works in any form on the internet, we would be grateful if you would provide us with the location address or website name. Please contact us at copyright@packt.com with a link to the material.

If you are interested in becoming an author: If there is a topic that you have expertise in and you are interested in either writing or contributing to a book, please visit authors.packtpub.com.

Share Your Thoughts

Once you've read *Serverless Machine Learning with Amazon Redshift ML*, we'd love to hear your thoughts! Scan the QR code below to go straight to the Amazon review page for this book and share your feedback.

https://packt.link/r/1-804-61928-0

Your review is important to us and the tech community and will help us make sure we're delivering excellent quality content.

Download a free PDF copy of this book

Thanks for purchasing this book!

Do you like to read on the go but are unable to carry your print books everywhere? Is your eBook purchase not compatible with the device of your choice?

Don't worry, now with every Packt book you get a DRM-free PDF version of that book at no cost.

Read anywhere, any place, on any device. Search, copy, and paste code from your favorite technical books directly into your application.

The perks don't stop there, you can get exclusive access to discounts, newsletters, and great free content in your inbox daily

Follow these simple steps to get the benefits:

1. Scan the QR code or visit the link below

https://packt.link/free-ebook/978-1-80461-928-5

2. Submit your proof of purchase
3. That's it! We'll send your free PDF and other benefits to your email directly

Part 1: Redshift Overview: Getting Started with Redshift Serverless and an Introduction to Machine Learning

Reaping the benefits of machine learning across an organization requires access to data, easy-to-use tools, and built-in algorithms that anyone can use no matter their level of experience with machine learning.

Part 1 shows how easy it is to get started with Amazon Redshift Serverless and Amazon Redshift ML without having to manage data warehouse infrastructure.

By the end of *Part 1*, you will know how to run queries using Query Editor v2 notebooks and different techniques for loading data into Amazon Redshift Serverless. You will then be introduced to machine learning and gain an understanding of how you can use machine learning in your data warehouse.

This part comprises the following chapters:

- *Chapter 1, Introduction to Amazon Redshift Serverless*
- *Chapter 2, Data Loading and Analytics on Redshift Serverless*
- *Chapter 3, Applying Machine Learning in Your Data Warehouse*

1

Introduction to Amazon Redshift Serverless

"Hey, what's a data warehouse?" John Doe, CEO and co-founder of Red.wines, a fictional specialty wine e-commerce company, asked Tathya Vishleshak*, the company's CTO. John, who owned a boutique winery, had teamed up with Tathya for the project. The company's success surged during the pandemic, driven by social media and the stay-at-home trend. John wanted detailed data analysis to align inventory and customer outreach. However, there was a problem – producing this analysis was slowing down their **online transaction processing** (**OLTP**) database.

"A data warehouse is like a big database where we store different data for a long time to find insights and make decisions," Tathya explained.

John had a concern, *"Sounds expensive; we're already paying for unused warehouse space. Can we afford it?"*

Tathya reassured him, *"You're right, but there are cloud data warehouses such as Amazon Redshift Serverless that let you pay as you use."*

Expanding on this, this chapter introduces data warehousing and Amazon Redshift. We'll cover Amazon Redshift Serverless basics, such as namespaces and workgroups, and guide you in creating a data warehouse. Amazon Redshift can gather data from various sources, mainly Amazon **Simple Storage Service** (**S3**).

As we go through this chapter, you'll learn about a crucial aspect of this, the AWS **Identity and Access Management** (**IAM**) role, needed for loading data from S3. This role connects to your Serverless namespace for smooth data transfer. You'll also learn how to load sample data and run queries using Amazon Redshift query editor. Our goal is to make it simple and actionable, so you're confident in navigating this journey.

> **Tathya Vishleshak**
>
> The phrase 'Tathya Vishleshak' can be loosely interpreted to reflect the concept of a data analyst in Sanskrit/Hindi. However, it's important to note that this is not a precise or established translation, but rather an attempt to convey a similar meaning based on the individual meanings of the words 'Tathya' and 'Vishleshak' in Sanskrit.

Additionally, Amazon Redshift is used to analyze structured and unstructured data in data warehouses, operational databases, and data lakes. It's employed for traditional data warehousing, business intelligence, real-time analytics, and machine learning/predictive analytics. Data analysts and developers use Redshift data with **machine learning** (**ML**) models for tasks such as predicting customer behavior. Amazon Redshift ML streamlines this process using familiar SQL commands.

The book delves into ML, explaining supervised and unsupervised training. You'll learn about problem-solving with binary classification, multi-class classification, and regression using real-world examples. You'll also discover how to create deep learning models and custom models with XGBoost, as well as use time series forecasting. The book also covers in-database and remote inferences using existing models, applying ML for predictive analytics, and operationalizing machine learning models.

The following topics will be covered in this chapter:

- What is Amazon Redshift?
- Getting started with Amazon Redshift Serverless
- Connecting to your data warehouse

This chapter requires a web browser and access to an AWS account.

What is Amazon Redshift?

Organizations churn out vast troves of customer data along with insights into these customers' interactions with the business. This data gets funneled into various applications and stashed away in disconnected systems. A conundrum arises when attempting to decipher these data silos – a formidable challenge that hampers the derivation of meaningful insights essential for organizational clarity. Adding to this complexity, security and performance considerations typically muzzle business analysts from accessing data within OLTP systems.

The hiccup is that intricate analytical queries weigh down OLTP databases, casting a shadow over their core operations. Here, the solution is the **data warehouse**, which is a central hub of curated data, used by business analysts and data scientists to make informed decisions by employing the business intelligence and machine learning tools at their disposal. These users make use of **Structured Query Language** (**SQL**) to derive insights from this data trove. From operational systems, application logs, and social media streams to the influx of IoT device-generated data, customers channel structured and semi-structured data into organizations' data warehouses, as depicted in *Figure 1.1*, showcasing the classic architecture of a conventional data warehouse.

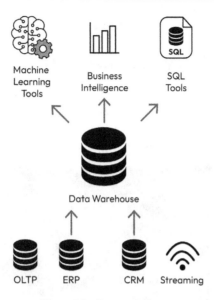

Figure 1.1 – Data warehouse

Here's where Amazon Redshift Serverless comes in. It's a key option within Amazon Redshift, a well-managed cloud data warehouse offered by **Amazon Web Services** (**AWS**). With cloud-based ease, Amazon Redshift Serverless lets you set up your data storage without infrastructure hassles or cost worries. You pay based on what you use for compute and storage.

Amazon Redshift Serverless goes beyond convenience, propelling modern data applications that seamlessly connect to the data lake. Enter the data lake – a structure that gathers all data strands under one roof, providing limitless space to store data at any scale, cost-effectively. Alongside other data repositories such as data warehouses, data lakes redefine how organizations handle data. And this is where it all comes together – the following diagram shows how Amazon Redshift Serverless injects SQL-powered queries into the data lake, driving a dynamic data flow:

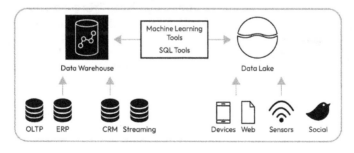

Figure 1.2 – Data lake and data warehouse

So, let's get started on creating our first data warehouse in the cloud!

Getting started with Amazon Redshift Serverless

You can create your data warehouse with Amazon Redshift Serverless using the **AWS Command-Line Interface (CLI)**, the **API**, **AWS CloudFormation templates**, or the **AWS console**. We are going to use the AWS console to create a Redshift Serverless data warehouse. Log in to your AWS console and search for `Redshift` in the top bar, as shown in *Figure 1.3*:

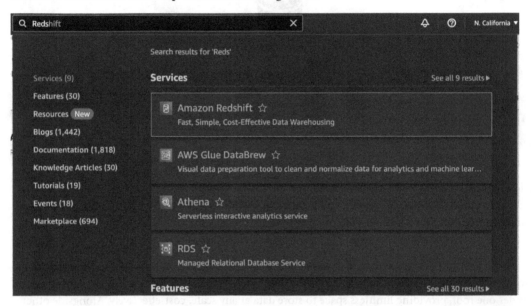

Figure 1.3 – AWS console page showing services filtered by our search for Redshift

Click on **Amazon Redshift**, which will take you to the home page for the Amazon Redshift console, as shown in *Figure 1.4*. To help get you started, Amazon provides free credit for first-time Redshift Serverless customers. So, let's start creating your trial data warehouse by clicking on **Try Amazon Redshift Serverless**. If you or your organization has tried Amazon Redshift Serverless before, you will have to pay for the service based on your usage:

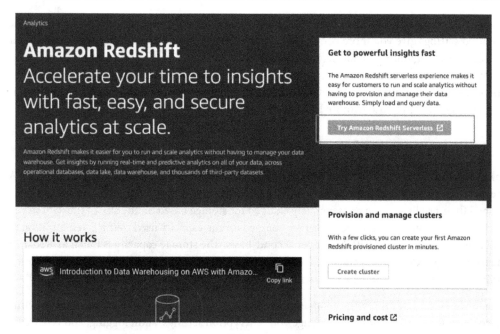

Figure 1.4 – Amazon Redshift service page in the AWS console

If you have free credit available, it will be indicated at the top of your screen, as in *Figure 1.5*:

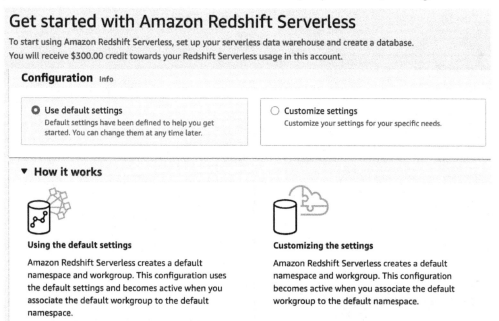

Figure 1.5 – AWS console showing the Redshift Serverless Get started page

You can either choose the defaults or use the customized settings to create your data warehouse. The customized settings give you more control, allowing you to specify many additional parameters for your compute configuration including the workgroup, data-related settings such as the namespace, and advanced security settings. We will use the customized settings, which will help us customize the namespace settings for our Serverless data warehouse. A namespace combined with a workgroup is what makes a data warehouse with Redshift Serverless, as we will now see in more detail.

What is a namespace?

Amazon Redshift Serverless provides a separation of storage and compute for a data warehouse. A **namespace** is a collection of all your data stored in the database such as your tables, views, database users, and their privileges. You are separately charged for storage based on the size of the data stored in your data warehouse. For compute, you are charged for the capacity used over a given duration in **Redshift processing hours** (RPU) on a per second-basis. The storage capacity is billed as **Redshift managed storage** (RMS) and is billed by GB/month. You can view https://aws.amazon.com/redshift/pricing/ for detailed pricing for your AWS Region.

As a data warehouse admin, you can change the name of your data warehouse namespace while creating the namespace. You can also change your encryption settings, audit logging, and AWS IAM permissions, as shown in *Figure 1.6*. The primary reason we are going to use customized settings is to associate an IAM role with the namespace:

Configuration Info

- ○ **Use default settings**
  Default settings have been defined to help you get started. You can change them at any time later.

- ◉ **Customize settings**
  Customize your settings for your specific needs.

Namespace Info

Namespace is a collection of database objects and users. Data properties include database name and password, permissions, and encryption and security.

Namespace name
This is a unique name that defines the namespace.

```
default
```

The name must be from 3-64 characters. Valid characters are a-z (lowercase only), 0-9 (numbers), and - (hyphen).

▼ **Database name and password**

Database name
The name of the first database in the Amazon Redshift Serverless environment.

```
dev
```

The name must be 1-64 alphanumeric characters (lowercase only), and it can't be a reserved word.

Admin user credentials
IAM credentials provided as your default admin user credentials. To add a new admin username and password, customize admin user credentials.

☐ **Customize admin user credentials**
To use the default IAM credentials, clear this option.

Figure 1.6 – Namespace configuration

AWS IAM allows you to specify which users or services can access other services and resources in AWS. We will use that role for loading data from **S3** and training a machine learning model with Redshift ML that accesses **Amazon SageMaker**.

If you have already created an IAM role earlier, you can associate with that IAM role. If you have not created an IAM role, do so now by selecting the **Manage IAM roles** option, as shown in *Figure 1.7*:

Figure 1.7 – Creating an IAM role and associating it via the AWS console

Then, select the **Create IAM role** option, as shown in *Figure 1.8*:

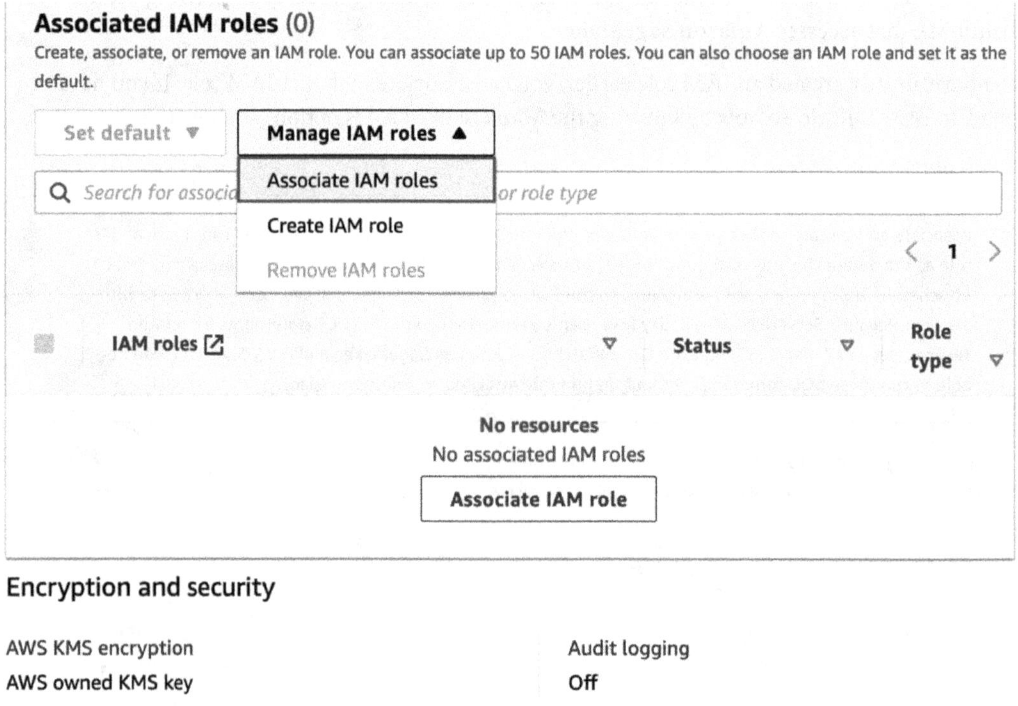

Associated IAM roles (0)

Create, associate, or remove an IAM role. You can associate up to 50 IAM roles. You can also choose an IAM role and set it as the default.

Set default ▼	Manage IAM roles ▲
	Associate IAM roles
Q Search for associa̶ or role type	Create IAM role
	Remove IAM roles

‹ 1 ›

	IAM roles ⤢	▽	Status	▽	Role type	▽

No resources
No associated IAM roles

Associate IAM role

Encryption and security

AWS KMS encryption   Audit logging
AWS owned KMS key   Off

Figure 1.8 – Selecting the "Create IAM role" option

You can then create a default IAM role and provide appropriate permissions to the IAM role to allow it to access S3 buckets, as shown in *Figure 1.9*:

Create the default IAM role   ✕

ⓘ Associate an IAM role so that your serverless endpoint can LOAD and
UNLOAD data. You can create an IAM role as the default for this configuration
that has the **AmazonRedshiftAllCommandsFullAccess** ☑ policy attached.
This policy includes permissions to run SQL commands to COPY, UNLOAD,
and query data with Amazon Redshift Serverless. This policy also grants
permissions to run SELECT statements for related services, such as Amazon
S3, Amazon CloudWatch logs, Amazon SageMaker, and AWS Glue. You won't
be able to run these SQL commands without an IAM role attached to your
namespace.

Specify an S3 bucket for the IAM role to access
To create a new bucket, visit S3 ☑

◯ No additional S3 bucket
Create the IAM role without specifying S3 buckets.

◉ Any S3 bucket
Allow users that have access to your Redshift Serverless data to also access any S3 bucket and its
contents in your AWS account.

◯ Specific S3 buckets
Specify one or more S3 buckets that the IAM role being created has permission to access.

Cancel	Create IAM role as default

Figure 1.9 – Granting S3 permissions to the IAM role

As shown in the preceding figure, select **Any S3 bucket** to enable Redshift to read data from and write
data to all S3 buckets you have created. Then, select **Create IAM role as default** to create the role and
set it as the default IAM role.

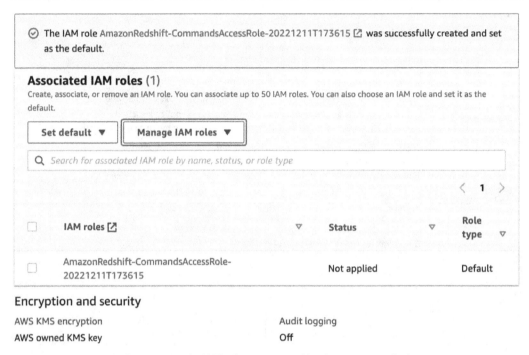

Figure 1.10 – An IAM role was created but is not yet applied

As shown in *Figure 1.10*, we created the IAM role and associated it with the namespace as a default role. Let's next proceed to create a workgroup, wherein we will set up the compute configuration for the data warehouse.

What is a workgroup?

As we discussed earlier, a namespace combined with a workgroup is what makes a Redshift Serverless data warehouse. A **workgroup** provides the compute resources required to process your data. It also provides the endpoint for you to connect to the warehouse. As an admin, you need to configure the compute settings such as the network and security configuration for the workgroup.

We will not do any customization at this time and simply select the default settings instead, including the VPC and associated subnets for the workgroup, as shown in the following screenshot:

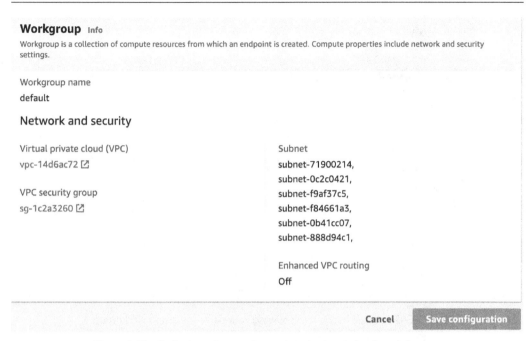

Figure 1.11 – Default settings and associated subnets for the workgroup

Click on the **Save configuration** button to create your Redshift Serverless instance, and your first data warehouse will be ready in a few minutes:

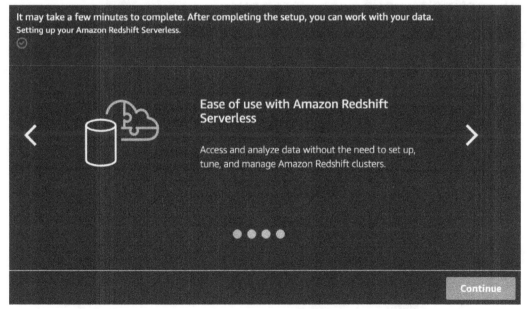

Figure 1.12 – Redshift Serverless creation progress

Once your data warehouse is ready, you will be redirected to your Serverless dashboard, as shown in *Figure 1.13*:

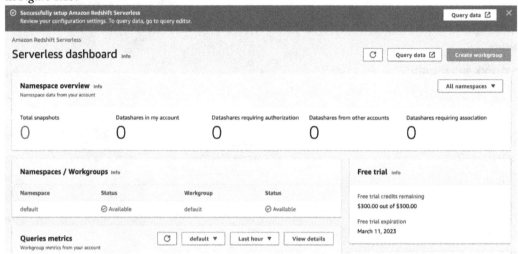

Figure 1.13 – Serverless dashboard showing your namespace and workgroup

Now that we have created our data warehouse, we will connect to the data warehouse, load some sample data, and run some queries.

Connecting to your data warehouse

Your data warehouse with Redshift Serverless is now ready. You can connect to your data warehouse using third-party tools via JDBC/ODBC/Python drivers. Other options include the **Data API** or the embedded **Redshift query editor v2**.

Using Amazon Redshift query editor v2

Now that your data warehouse is ready; let's navigate to the query editor to load some sample data and run some queries. Select the **Query data** option from your dashboard, as shown in *Figure 1.13*, and you will be navigated to the query editor, as shown in *Figure 1.14*.

Figure 1.14 – Query editor

In the Redshift query editor v2 console, on the left pane, you will see the data warehouses, such as the **Serverless:default** workgroup, that you have access to. Click on the workgroup (**Serverless:default**) to connect to the data warehouse.

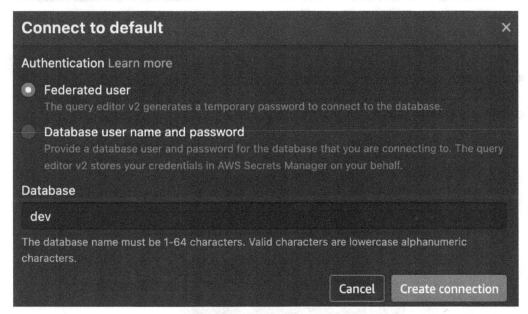

Figure 1.15 – Creating a connection to your workgroup

As shown in the preceding screenshot, select **Federated user** if you did not specify any database credentials while creating the namespace, and then click **Create connection**. You can leave the database name as **dev**. You will be prompted to create a connection only when connecting to the data warehouse for the first time. If you have created the connection, you will be connected automatically when you click on the workgroup. Once you are connected, you will see the databases in the navigator, as shown in *Figure 1.16*:

Figure 1.16 – List of databases

Since we just created our data warehouse for the first time, there is no data present in it, so let's load some sample data into the data warehouse now.

Loading sample data

On the left pane, click on the **sample_data_dev** database to expand the available database:

Figure 1.17 – The Redshift query editor v2 navigator that shows the sample data available

As you can see from the preceding screenshot, three sample datasets are available for you to load into your data warehouse. Click on the icon showing the folder with an arrow located to the right of your chosen sample data notebook to load and open it, as shown in *Figure 1.18*:

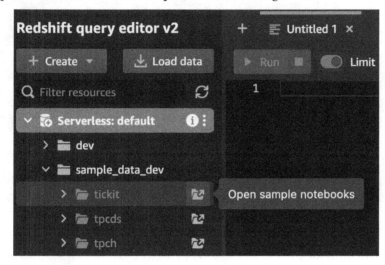

Figure 1.18 – List of sample databases

You will be prompted to create your sample database. Click on **Create** to get started, as shown in *Figure 1.19*:

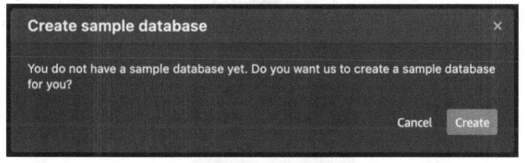

Figure 1.19 – Creating a sample database

The sample data will be loaded in a few seconds and presented in a notebook with SQL queries for the dataset that you can explore, as shown in *Figure 1.20*:

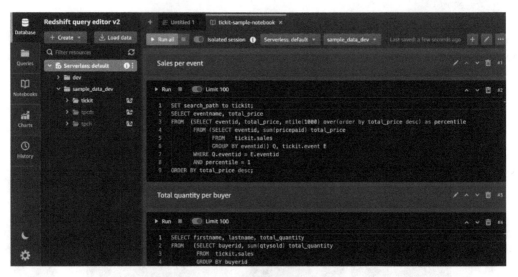

Figure 1.20 – Notebook with sample queries for the tickit database

You can expand the navigation tree on the left side of the query editor to view schemas and database objects, such as tables and views in your schema, as shown in *Figure 1.21*.

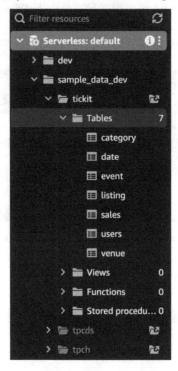

Figure 1.21 – Expanding the navigation tree to view schemas and database objects

You can click on a table to view the table definitions, as shown in *Figure 1.22*:

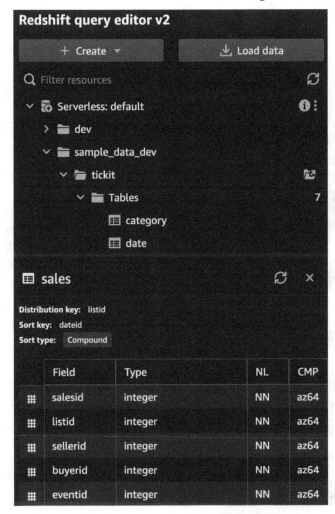

Figure 1.22 – Table definitions

Right-clicking on a table provides additional **Select table**, **Show table definition**, and **Delete** options, as shown in *Figure 1.23*:

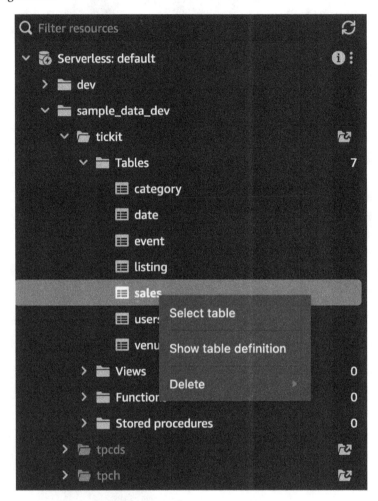

Figure 1.23 – Right-clicking on a table to view more options

You can click **Run all,** as shown in *Figure 1.24,* to run all the queries in the sample notebook. The query editor provides a notebook interface to add annotation, and SQL cells organize your queries in a single document. You can use annotations for documentation purposes.

Figure 1.24 – The "Run all" option

You will see the results of your queries for each cell. You can download the results as JSON or CSV files to your desktop, as shown in *Figure 1.25*:

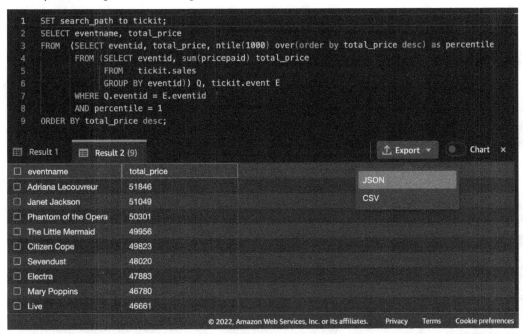

Figure 1.25 – Options to download query results

Let's author our first query.

Running your first query

We want to find out the top 10 events by sales in the `tickit` database. We will run the following SQL statement in the data warehouse:

```
SELECT eventname, total_price
FROM  (SELECT eventid, total_price, ntile(1000) over(order by total_
price desc) as percentile
      FROM (SELECT eventid, sum(pricepaid) total_price
            FROM   tickit.sales
            GROUP BY eventid)) Q, tickit.event E
      WHERE Q.eventid = E.eventid
      AND percentile = 1
ORDER BY total_price desc
limit 10;
```

In the query editor, add a new query by clicking on the + sign and selecting **Editor** from the menu that appears. If you wanted to create a new notebook, you could click on **Notebook** instead, as shown in *Figure 1.26*:

Figure 1.26 – Creating a new query

Now, type the preceding SQL query in the editor and then click on **Run**. You will get the results as shown in the following screenshot:

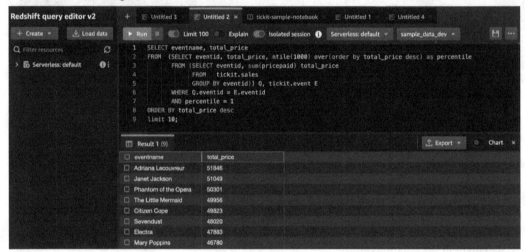

Figure 1.27 – Query with results

As the saying goes, *"A picture is worth a thousand words,"* and query editor allows you to visualize the results to gain faster insight. You can create a chart easily by clicking on the **Chart** option and then selecting the chart you want. Let's select a scatter plot, as shown in *Figure 1.28*:

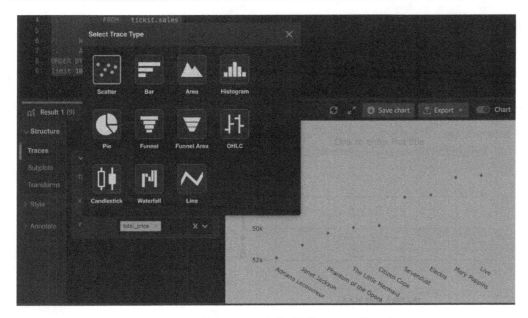

Figure 1.28 – Using charts in Redshift query editor v2

You can add a chart name and notations for the X and Y axes and export the chart as PNG or JPG to put in your presentation or to share with your business partners:

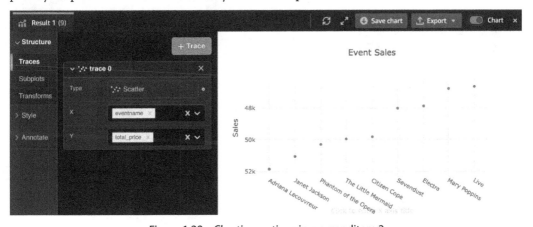

Figure 1.29 – Charting options in query editor v2

As you have now seen, you can use Redshift query editor v2 to create your own database, create tables, load data, and run and author queries and notebooks. You can share your queries and notebooks with your team members.

Summary

In this chapter, you learned about cloud data warehouses and Amazon Redshift Serverless. You created your first data warehouse powered by Redshift Serverless and loaded some sample data using the query editor. You also learned how to use the query editor to run queries and visualize data to produce insights.

In *Chapter 2*, you will learn the best techniques for loading data and performing analytics in your Amazon Redshift Serverless data warehouse.

2

Data Loading and Analytics on Redshift Serverless

In the previous chapter, we introduced you to **Amazon Redshift Serverless** and demonstrated how to create a serverless endpoint from the Amazon Redshift console. We also explained how to connect and query your data warehouse using **Amazon Redshift query editor v** In this chapter, we will dive deeper into the different ways you can load data into your Amazon Redshift Serverless data warehouse.

We will cover three main topics in this chapter to help you load your data efficiently into Redshift Serverless. First, we will demonstrate how to load data using Amazon Redshift query editor v where you will learn how to load data from your Amazon S3 bucket and local data file onto your computer using the GUI.

Next, we will explore the COPY command in detail, and you will learn how to load a file by writing a COPY command to load the data. We will cover everything you need to know to use this command effectively and load your data smoothly into Redshift Serverless.

Finally, we will cover the built-in native API interface to access and load data into your Redshift Serverless endpoint using Jupyter Notebook. We will guide you through the process of setting up and using the **Redshift Data API**.

The topics are as follows:

- Data loading using Amazon Redshift query editor v
- Data loading from Amazon S3 using the COPY command
- Data loading using the Redshift Data API

The goal of this chapter is to equip you with the knowledge and skills to load data into Amazon Redshift Serverless using different mechanisms. By the end of this chapter, you will be able to load data quickly and efficiently into Redshift Serverless using the methods covered in this chapter, which will enable you to perform analytics on your data and extract valuable insights.

Technical requirements

This chapter requires a web browser and access to the following:

- An AWS account
- Amazon Redshift
- Amazon Redshift Query Editor v2
- Amazon SageMaker for Jupyter Notebook

The code snippets in this chapter are available in this book's GitHub repository at `https://github.com/PacktPublishing/Serverless-Machine-Learning-with-Amazon-Redshift/tree/main/CodeFiles/chapter2`.

The data files used in this chapter can be found in this book's GitHub repository: `https://github.com/PacktPublishing/Serverless-Machine-Learning-with-Amazon-Redshift/tree/main/DataFiles/chapter2`.

Data loading using Amazon Redshift Query Editor v2

Query Editor v2 supports different database actions, including **data definition language (DDL)**, to create schema and tables and load data from data files with just a click of a button. Let's take a look at how you can carry out these tasks to enable easy analytics on your data warehouse. Log in to your AWS console, navigate to your Amazon Redshift Serverless endpoint, and select **Query data**. This will open **Redshift query editor v2** in a new tab. Using the steps we followed in *Chapter 1*, log in to your database and perform the tasks outlined in the following subsections.

Creating tables

Query editor v2 provides a wizard to execute the DDL commands shown in *Figure 2.1*. Let's create a new schema named `chapter2` first:

1. Click on **Create** and select **Schema**, as shown here.

Figure 2.1 – The creation wizard

Ensure that your **Cluster or workgroup** and **Database** parameters are correctly populated. If not, then select the correct values from the dropdowns. Give a suitable name for your schema; we will name it `chapter2`.

2.   Then, click on **Create schema**, as shown in *Figure 2.2*:

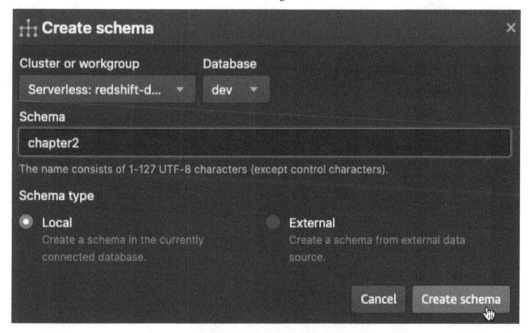

Figure 2.2 – Create schema

Once you have your schema created, navigate to the **Create** drop-down button and click on **Table**. This will open up the **Create table** wizard. Select the appropriate values for your workgroup and database, and enter `chapter2` in the **Schema** field. Give your table the name `customer`. With Query Editor v2, you can either enter the column names and their data type manually, or you can use the data file to automatically infer the column names and their data type.

Let's create a table with a data file. We will use `customer.csv`, which is available in this book's GitHub repository at `https://github.com/PacktPublishing/Serverless-Machine-Learning-with-Amazon-Redshift/tree/main/DataFiles/chapter2`. You can download this file locally to create the table using the wizard.

The file contains a subset of the data from the `TPC-H` dataset, available in this book's GitHub repository: `https://github.com/awslabs/amazon-redshift-utils/tree/master/src/CloudDataWarehouseBenchmark/Cloud-DWB-Derived-from-TPCH`.

On the **Create table** wizard, click on **Load from CSV** under the **Columns** tab, and provide a path to the CSV file. Once the file is selected, the schema will be inferred and automatically populated from the file, as shown in *Figure 2.3*. Optionally, you can modify the schema in the **Column name**, **Data type**, and **Encoding** fields, and under **Column options**, you can select different options such as the following:

- Choose a default value for the column.
- Optionally, you can turn on **Automatically increment** if you want the column values to increment. If you enable this option, only then can you specify a value for **Auto increment seed** and **Auto increment step**.
- Enter a size value for the column.
- You also have the option to define constraints such as **Not NULL**, **Primary key**, and **Unique key**.

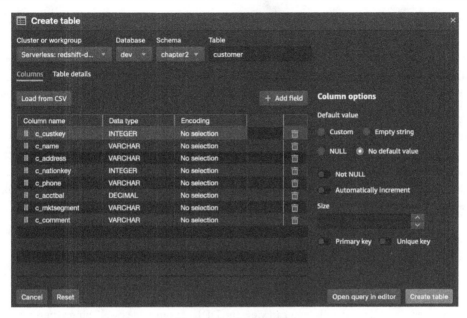

Figure 2.3 – Create table

Additionally, as shown in *Figure 2.4*, under the **Table details** tab, you can optionally set the table properties, such as **Distribution key**, **Distribution style**, **Sort key**, and **Sort type**. When these options are not set, Redshift will pick default settings for you, which are **Auto Distribution Key** and **Auto Sort Key**.

Figure 2.4 – Table details

Amazon Redshift distributes data in a table according to the table's distribution style (`DISTSTYLE`). The data rows are distributed within each compute node according to the number of slices. When you run a query against the table, all the slices of the compute node process the rows that are assigned in parallel. As a best practice (`https://docs.aws.amazon.com/redshift/latest/dg/c_best-practices-best-dist-key.html`), select a table's `DISTSTYLE` parameter to ensure even distribution of the data or use automatic distribution.

Amazon Redshift orders data within each slice using the table's sort key. Amazon Redshift also enables you to define a table with compound sort keys, interleaved sort keys, or no sort keys. As a best practice (`https://docs.aws.amazon.com/redshift/latest/dg/c_best-practices-sort-key.html`), define the sort keys and style according to your data access pattern. Having a proper sort key defined on a table can hugely improve your query performance.

Lastly, under **Other options** you can select the following:

- Whether to include your table in automated and manual snapshots
- Whether to create a session-based temporary table instead of a permanent database table

Once you have entered all the details, you can view the DDL of your table by clicking **Open query in editor**. You can use this later or even share it with other users.

Now, let's create our table by clicking on the **Create table** button (*Figure 2.4*).

As you can see, it is easy for any data scientist, analyst, or user to use this wizard to create database objects (such as tables) without having to write DDL and enter each column's data type and its length.

Let's now work on loading data in the customer table. Query Editor v2 enables you to load data from Amazon S3 or the local file on your computer. Please note that, at the time of writing, the option to load a local file currently supports only CSV files with a maximum size of 5 MB.

Loading data from Amazon S3

Query editor v2 enables you to load data from Amazon S3 buckets into an existing Redshift table.

The **Load data** wizard populates data from Amazon S3 by generating the `COPY` command, which really makes it easier for a data analyst or data scientist, as they don't have to remember the intricacies of the `COPY` command. You can load data from various file formats supported by the `COPY` command, such as CSV, JSON, Parquet, Avro, and Orc. Refer to this link for all the supported data formats: `https://docs.aws.amazon.com/redshift/latest/dg/copy-parameters-data-format.html#copy-format`.

Let's look at loading data using the **Load data** wizard. We will load the data into our customer table from our data file (`customer.csv`), which is stored in the following Amazon S3 location: `s3://packt-serverless-ml-redshift/chapter02/customer.csv`.

Note that if you want to use your own Amazon S3 bucket to load the data, then download the data file from the GitHub location mentioned in the *Technical requirements* section.

To download a data file from GitHub, navigate to your repository, select the file, right-click the **View raw** button at the top of the file, select **Save Link As…** (as shown in the following screenshot), choose the location on your computer where you want to save the file, and select **Save**:

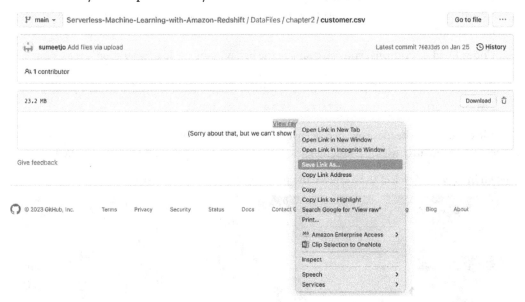

Figure 2.5 – Saving the data file

On Query Editor v2, click on **Load data**, which opens up the data load wizard.

Under **Data source**, select the **Load from S3** radio button. You can browse the S3 bucket in your account to select the data file or a folder that you want to load, or you can select a manifest file. For this exercise, paste the aforementioned S3 file location.

If the data file is in a different region than your Amazon Redshift Serverless, you can select the source region from the S3 file location dropdown. The wizard provides different options if you want to load a Parquet file. Then, select an option from **File format**, or under **File** options, you can select **Delimiter** if your data is delimited by a different character. If your file is compressed, then you can select the appropriate compression from the dropdown, such as **gzip**, **lzop**, **zstd**, or **bzip2**.

Under **Advanced settings**, note that there are two options, **Data conversion parameters** and **Load operations**:

- Under the **Data conversion parameters** option, you can handle explicit data conversion settings – for example, a time format (TIMEFORMAT) as `MM.DD.YYYY HH:MI:SS`. Refer to this documentation link for a full list of parameters: `https://docs.aws.amazon.com/redshift/latest/dg/copy-parameters-data-conversion.html#copy-timeformat`.

- Under **Load operations**, you can manage the behavior of the load operation – for example, the number of rows for compression analysis (COMPROWS) as `1,000,000`. Refer to this documentation for a full list of options: `https://docs.aws.amazon.com/redshift/latest/dg/copy-parameters-data-load.html`.

As our file contains the header row, please ensure that under **Advanced settings | Data conversion parameters | Frequently used parameters**, the **Ignore header rows (as 1)** option is checked.

As shown in *Figure 2.6*, select the **Target table** parameters and **IAM role** to load the data:

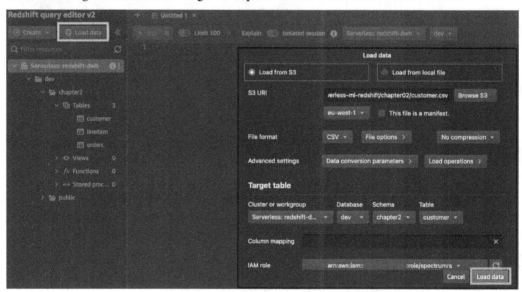

Figure 2.6 – Load data

Once you click on **Load data**, Query Editor v2 will generate the COPY command in the editor and start loading by running the COPY statement.

Now that we have loaded our data, let's quickly verify the load and check the data by querying the table, as shown here:

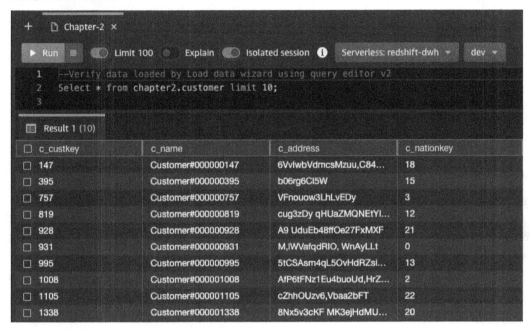

Figure 2.7 – Querying the data

Query Editor v2 enables you to save your queries in the editor for later use. You can do so by clicking on the **Save** button and providing a name for the saved query. For example, if you want to reuse the preceding load data query (the COPY command) in the future and, let's say, the target table is the same but the data location on Amazon S3 is different, then you can easily modify this query and load the data quickly. Alternatively, you can even parameterize the query to pass, for example, an S3 location as ${s3_location}, as shown in *Figure 2.8*:

```
1   COPY dev.chapter2.customer
2   FROM ${s3_location}
3   IAM_ROLE 'arn:aws:iam::         :role/spectrumrs'
4   FORMAT AS CSV DELIMITER ','
5   QUOTE '"'
6   IGNOREHEADER 1
7   REGION AS 'us-west-2'
```

Figure 2.8 – Saving the query

> **Sharing queries**
>
> With Query Editor v2, you can share your saved queries with your team. This way, many users can collaborate and share the same query. Internally, Query Editor manages the query versions, so you can track the changes as well. To learn more about this, refer to this AWS documentation: `https://docs.aws.amazon.com/redshift/latest/mgmt/query-editor-v2-team.html#query-editor-v2-query-share`.

Now that we have covered how Query Editor v2 enables users to easily create database objects and load data using the UI interface with a click of a few buttons, let us dive into Amazon Redshift's COPY command to load the data into your data warehouse.

Loading data from a local drive

Query Editor v2 enables users to load data from a local file on their computer and perform analysis on it quickly. Often, database users such as data analysts or data scientists have data files on their local computer that they want to load quickly into a Redshift table, without moving the file into a remote location such as Amazon S3.

In order to load the data from a local file, Query Editor v2 requires a staging Amazon S3 bucket in your account. If it is not configured, then you will see an error similar to the one seen in the following screenshot:

Figure 2.9 – An error message

To avoid the preceding error, users must do the following configuration:

1. The account users must be configured with the proper permissions, as follows. Attach the following policy to your Redshift Serverless IAM role. Replace the resource names as highlighted:

    ```
    {
        "Version": "2012-10-17",
        "Statement": [
            {
    ```

```
                    "Effect": "Allow",
                    "Action": [
                        "s3:ListBucket",
                        "s3:GetBucketLocation"
                    ],
                    "Resource": [
                        "arn:aws:s3:::<staging-bucket-name>"
                    ]
                },
                {

                    "Effect": "Allow",
                    "Action": [
                        "s3:PutObject",
                        "s3:GetObject",
                        "s3:DeleteObject"
                    ],
                    "Resource": [
                        "arn:aws:s3:::<staging-bucket-name>[/<optional-
prefix>]/${aws:userid}/*"
                    ]
                }
            ]
}
```

2. Your administrator must configure the common Amazon S3 bucket in the **Account settings** window, as shown here:

    I. Click on the settings icon ( ) and select **Account settings**, as shown in the following screenshot:

Figure 2.10 – Account settings

    II. In the **Account settings** window, under **General settings** | **S3 bucket** | **S3 URI**, enter the URI of the S3 bucket that will be used for staging during the local file load, and then click on **Save**. Ensure that your IAM role has permission to read and write on the S3 bucket:

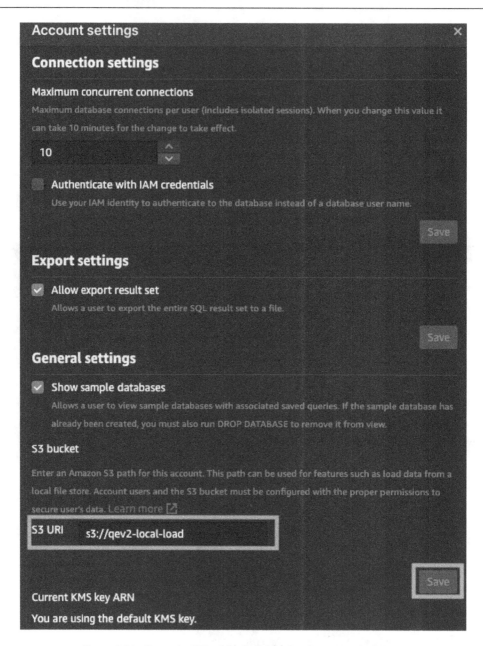

Figure 2.11 – Enter the URI of the S3 bucket under General settings

Refer to this documentation for complete information:

```
https://docs.aws.amazon.com/redshift/latest/mgmt/query-editor-v2-
loading.html#query-editor-v2-loading-data-local
```

Creating a table and loading data from a local CSV file

Let's create a new table. Navigate to Query Editor v2 and create a supplier table using the following DDL command:

```
CREATE TABLE chapter2.supplier (
    s_suppkey integer NOT NULL ENCODE raw distkey,
    s_name character(25) NOT NULL ENCODE lzo,
    s_address character varying(40) NOT NULL ENCODE lzo,
    s_nationkey integer NOT NULL ENCODE az64,
    s_phone character(15) NOT NULL ENCODE lzo,
    s_acctbal numeric(12, 2) NOT NULL ENCODE az64,
    s_comment character varying(101) NOT NULL ENCODE lzo,
    PRIMARY KEY (s_suppkey)
) DISTSTYLE KEY
SORTKEY (s_suppkey);
```

We will load the data into our supplier table from our data file (`supplier.csv`), which is stored in the following GitHub location: `https://github.com/PacktPublishing/Serverless-Machine-Learning-with-Amazon-Redshift/blob/main/DataFiles/chapter2/supplier.csv`.

To download the file on your local computer, right-click on **Raw** and click on **Save Link as**.

In order to load data into the supplier table from Query Editor v2, click on **Load data**, which opens up the data load wizard. Under **Data source**, select the **Load from local file** radio button. Click on **Browse** and select the `supplier.csv` file from your local drive. Under the **Target table** options, set **Schema** as **chapter2** and **Table** as **supplier**. Click on **Load data** to start the load:

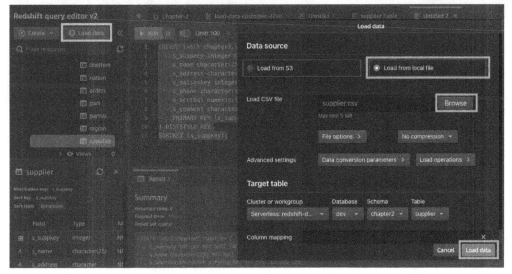

Figure 2.12 – The Load data wizard

Once the data is loaded successfully, you would see a message like the following:

Figure 2.13 – The message after successfully loading the data

Verify the data load by running the following SQL query:

```
select * from chapter2.supplier;
```

You should be able to see 100 rows loaded from the file:

▶ Run ■ ⬤ Limit 100 ⬤ Explain ⬤ Isolated session ⓘ Serverless: redshift-dwh ▾ dev ▾

```
1   select * from chapter2.supplier;
```

⊞ Result 1 (100)						⬆ Export ▾
☐ s_suppkey	s_name	s_address	s_nationkey	s_phone	s_acctbal	s_comment
☐ 1726	Supplier#000001726	TeRY7TtTH24sEword7yA...	3	13-630-597-4070	-751.93	ly alongside of the daringl...
☐ 1775	Supplier#000001775	a6Rpe337dA eJQl9RJZe...	12	22-193-554-4479	923.43	ously bold accounts. fluffi...
☐ 1889	Supplier#000001889	eK1A7NhiGccTJw78wxR T	19	29-122-640-8760	264.01	olites believe blithely fluffi...
☐ 2017	Supplier#000002017	5X3rJUk8 SA0E6RokqW...	1	11-940-342-7501	6990.21	y final deposits hang. bilt...
☐ 2082	Supplier#000002082	7v7kwZySU297XEFptPD...	16	26-179-265-4597	-664.94	dolites. carefully final acc...
☐ 2102	Supplier#000002102	1kuyUn5q6czLOGB60fAV...	11	21-367-198-9930	7910.16	accounts after the blithely
☐ 2540	Supplier#000002540	fpd9A3WsslyUkKypgdpV...	20	30-301-816-1967	1958.59	ly blithely regular request...
☐ 2615	Supplier#000002615	MIGmjZy93D	20	30-285-851-2800	1047.61	. pending packages haggl...
☐ 2652	Supplier#000002652	fAzRSsQ6VnxRSosL7aO...	4	14-882-858-2723	4366.99	requests. carefully even I...
☐ 2757	Supplier#000002757	CaERkgKvPpr4LVD7voAs...	9	19-959-961-5223	1085.89	ggle blithely. regular acco...
☐ 2769	Supplier#000002769	7QXYQHHuMvIkDPVwCi...	13	23-312-807-5269	7704.51	ackages cajole carefully. ...
☐ 2791	Supplier#000002791	qN7ZlkL2KHGHsX	10	20-951-419-1791	6468.16	across the deposits. blithely
☐ 2820	Supplier#000002820	Fmb9hufE41cerFFf2DiGF...	7	17-358-908-6872	6475.21	Customer Complaintshe ...
☐ 2953	Supplier#000002953	TepT5BpdFsnZT grAXI7,x...	16	26-118-226-8835	3955.99	usly final requests integra...
☐ 3091	Supplier#000003091	JNPFRBSLdD4DksRyryH...	14	24-781-138-5146	3530.16	. platelets sleep carefully i...

Figure 2.14 – Data load verification

We have now successfully loaded our data from the Query Editor v2 **Load data** wizard, using files from an Amazon S3 bucket and your local computer. Let's look into Amazon Redshift's COPY command in the next section.

Data loading from Amazon S3 using the COPY command

Data warehouses are typically designed to ingest and store huge volumes of data, and one of the key aspects of any analytical process is to ingest such huge volumes in the most efficient way. Loading such huge data can take a long time as well as consume a lot of compute resources. As pointed out earlier, there are several ways to load data in your Redshift Serverless data warehouse, and one of the fastest and most scalable methods is the COPY command.

The COPY command loads your data in parallel from files, taking advantage of Redshift's **massively parallel processing** (**MPP**) architecture. It can load data from Amazon S3, Amazon EMR, Amazon DynamoDB, or text files on remote hosts (SSH). It is the most efficient way to load a table in your Redshift data warehouse. With proper IAM policies, you can securely control who can access and load data in your database.

In the earlier section, we saw how Query Editor v2 generates the COPY command to load data from the wizard. In this section, we will dive deep and talk about how you can write the COPY command and load data from Amazon S3, and what some of the best practices are.

Let's take a look at the COPY command to load data into your Redshift data warehouse:

```
COPY table-name
[ column-list ]
FROM data_source
authorization
[ [ FORMAT ] [ AS ] data_format ]
[ parameter [ argument ] [, ... ] ]
```

The COPY command requires three parameters:

- table-name: The target table name existing in the database (persistent or temporary)
- data_source: The data source location (such as the S3 bucket)
- authorization: The authentication method (for example, the IAM role)

By default, the COPY command source data format is expected to be in character-delimited UTF-8 text files, with a pipe character (|) as the default delimiter. If your source data is in another format, you can pass it as a parameter to specify the data format. Amazon Redshift supports different data formats, such as fixed-width text files, character-delimited files, CSV, JSON, Parquet, and Avro.

Additionally, the COPY command provides optional parameters to handle data conversion such as the data format, null, and encoding. To get the latest details, refer to this AWS documentation: https://docs.aws.amazon.com/redshift/latest/dg/r_COPY.html#r_COPY-syntax.

Loading data from a Parquet file

In the earlier section, we worked on loading a CSV file into the customer table in our database. For this exercise, let's try to load a columnar data format file such as Parquet. We will be using a subset of TPC-H data, which may be found here: https://github.com/awslabs/amazon-redshift-utils/tree/master/src/CloudDataWarehouseBenchmark/Cloud-DWB-Derived-from-TPCH/3TB.

The TPC is an organization focused on developing data benchmark standards. You may read more about TPC here: https://www.tpc.org/default5.asp.

The modified data (`lineitem.parquet`) is available on GitHub: `https://github.com/PacktPublishing/Serverless-Machine-Learning-with-Amazon-Redshift/tree/main/DataFiles/chapter2`.

The data needed for the COPY command is available here: `s3://packt-serverless-ml-redshift/chapter02/lineitem.parquet`.

This file contains approximately 6 million rows and is around 200 MB in size:

1. Let's first start by creating a table named `lineitem` in the `chapter2` schema:

```
-- Create lineitem table
CREATE TABLE chapter2.lineitem
(l_orderkey       bigint,
 l_partkey        bigint,
 l_suppkey        integer,
 l_linenumber     integer,
 l_quantity       numeric(12,2),
 l_extendedprice  numeric(12,2),
 l_discount       numeric(12,2),
 l_tax            numeric(12,2),
 l_returnflag     character(1),
 l_linestatus     character(1),
 l_shipdate       date,
 l_commitdate     date,
 l_receiptdate    date,
 l_shipinstruct   character(25),
 l_shipmode       character(10),
 l_comment        varchar(44))
distkey(l_orderkey) compound sortkey(l_orderkey,l_shipdate);
```

2. Now, let's load the data using the COPY command from the `lineitem.parquet` file:

```
COPY chapter2.lineitem
FROM 's3://packt-serverless-ml-redshift/chapter02/lineitem.parquet'
IAM_ROLE default
FORMAT AS PARQUET;
```

Now that we have loaded our data, let's quickly verify the load and check the data by querying the table, as shown in the following screenshot:

	c_custkey	c_name	c_address	c_nationkey	c_phone	c_acctbal
☐	44	Customer#000000044	OI,dOSPwDu4jo4x,,P85E...	16	26-190-260-5375	7315.94
☐	251	Customer#000000251	Z9fdQmv07C3k hxwt9nc...	13	23-975-623-5949	9585.32
☐	381	Customer#000000381	w3zVseYDbjBbzLld	5	15-860-208-7093	9931.71
☐	469	Customer#000000469	JWOULMa5Qtt	12	22-406-988-6460	6343.64
☐	520	Customer#000000520	yaOGc9Ve92Bi4F6e0Gch...	3	13-612-111-7765	8315.09
☐	601	Customer#000000601	P3 Dv,6yIITNmL9yt6NUZ...	1	11-104-635-9839	9768.21
☐	618	Customer#000000618	9O4fhgteQdyFvCkrFm	0	10-675-573-1877	-932.38
☐	740	Customer#000000740	FCerGpsfsWAsBrQTyqdz...	10	20-215-156-3727	1733.76
☐	971	Customer#000000971	z29DUY Utsi6mWKI	1	11-256-718-6928	3914.88
☐	1053	Customer#000001053	wDJTteyausmZswQAFQot	16	26-400-312-6496	-473.85

Figure 2.15 – The query table

In this section, we discussed how the COPY command helps load your data in different formats, such as CSV, Parquet, and JSON, from Amazon S3 buckets. Let's see how you can automate the COPY command to load the data as soon as it is available in an Amazon S3 bucket. The next section on automating a COPY job is currently in public preview at the time of writing.

Automating file ingestion with a COPY job

In your data warehouse, data is continuously ingested from Amazon S3. Previously, you wrote custom code externally or locally to achieve this continuous ingestion of data with scheduling tools. With Amazon Redshift's auto-copy feature, users can easily automate data ingestion from Amazon S3 to Amazon Redshift. To achieve this, you will write a simple SQL command to create a COPY job (https://docs.aws.amazon.com/redshift/latest/dg/r_COPY-JOB.html), and the COPY command will trigger automatically as and when it detects new files in the source Amazon S3 path. This will ensure that users have the latest data for processing available shortly after it lands in the S3 path, without having to build an external custom framework.

To get started, you can set up a COPY job, as shown here, or modify the existing COPY command by adding the JOB CREATE parameter:

```
COPY <table-name>
FROM 's3://<s3-object-path>'
[COPY PARAMETERS...]
JOB CREATE <job-name> [AUTO ON | OFF];
```

Let's break this down:

- `job-name` is the name of the job
- `AUTO ON | OFF` indicates whether the data from Amazon S3 has loaded automatically into an Amazon Redshift table

As you can see, the `COPY` job is an extension of the `COPY` command, and auto-ingestion of `COPY` jobs is enabled by default.

If you want to run a `COPY` job, you can do so by running the following command:

```
COPY JOB RUN job-name
```

For the latest details, refer to this AWS documentation: `https://docs.aws.amazon.com/redshift/latest/dg/loading-data-copy-job.html`.

Best practices for the COPY command

The following best practices will help you get the most out of the `COPY` command:

- Make the most of parallel processing by splitting data into multiple compressed files or by defining distribution keys on your target tables, as we did in our example.
- Use a single `COPY` command to load data from multiple files. If you use multiple concurrent `COPY` commands to load the same target table from multiple files, then the load is done serially, which is much slower than a single `COPY` command.
- If your data file contains an uneven or mismatched number of fields, then provide the list of columns as comma-separated values.
- When you want to load a single target table from multiple data files and your data files have a similar structure but different naming conventions, or are in different folders in an Amazon S3 bucket, then use a manifest file. You can supply the full path of the files to be loaded in a JSON-formatted text file. The following is the syntax to use a manifest file:

```
copy <table_name> from 's3://<bucket_name>/<manifest_file>'
authorization
manifest;
```

- For a `COPY` job, use unique filenames for each file that you want to load. If a file is already processed and any changes are done after that, then the `COPY` job will not process the file, so remember to rename the updated file.

So far, we have seen two approaches to data loading in your Amazon Redshift data warehouse – using the Query Editor v2 wizard and writing an individual `COPY` command to trigger ad hoc data loading. Let us now look into how you can use an AWS SDK to load data using the Redshift Data API.

Data loading using the Redshift Data API

The Amazon Redshift Data API is a built-in native API interface to access your Amazon Redshift database without configuring any **Java Database Connectivity (JDBC)** or **Open Database Connectivity (ODBC)** drivers. You can ingest or query data with a simple API endpoint without managing a persistent connection. The Data API provides a secure way to access your database by using either IAM temporary credentials or AWS Secrets Manager. It provides a secure HTTP endpoint to run SQL statements asynchronously, meaning you can retrieve your results later. By default, your query results are stored for 24 hours. The Redshift Data API integrates seamlessly with different AWS SDKs, such as Python, Go, Java, Node.js, PHP, Ruby, and C++. You can also integrate the API with AWS Glue for an ETL data pipeline or use it with AWS Lambda to invoke different SQL statements.

There are many use cases where you can utilize the Redshift Data API, such as ETL orchestration with AWS Step Functions, web service-based applications, event-driven applications, and accessing your Amazon Redshift database using Jupyter notebooks. If you want to just run an individual SQL statement, then you can use the **AWS Command-Line Interface (AWS CLI)** or any programming language. The following is an example of executing a single SQL statement in Amazon Redshift Serverless from the AWS CLI:

```
aws redshift-data execute-statement
--WorkgroupName redshift-workgroup-name
--database dev
--sql 'select * from redshift_table';
```

Note that, for Redshift Serverless, you only need to provide the workgroup name and database name. Temporary user credentials are pulled from IAM authorization. For Redshift Serverless, add the following permission in the IAM policy attached to your cluster IAM role to access the Redshift Data API:

```
redshift-serverless:GetCredentials
```

In order to showcase how you can ingest data using the Redshift Data API, we will carry out the following steps using Jupyter Notebook. Let's create a notebook instance in our AWS account.

On the console home page, search for `Amazon SageMaker`. Click on the hamburger icon (☰) in the top-left corner, then **Notebook**, and then **Notebook instances**. Click on **Create notebook instance** and provide the necessary input. Once the notebook instance is in service, click on **Open Jupyter**.

The following screenshot shows a created notebook instance:

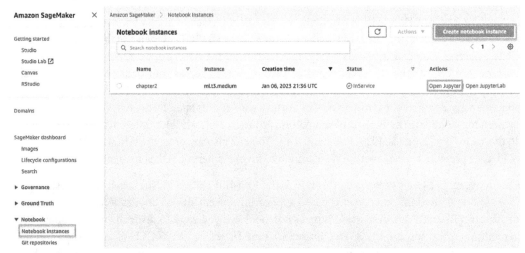

Figure 2.16 – Creating a notebook instance

The Jupyter notebook for this exercise is available at this GitHub location: `https://github.com/PacktPublishing/Serverless-Machine-Learning-with-Amazon-Redshift/blob/main/Chapter2.ipynb`. Download this notebook to your local machine and save it in a folder.

The data (`orders.parquet`) for this exercise is available on GitHub at `https://github.com/PacktPublishing/Serverless-Machine-Learning-with-Amazon-Redshift/tree/main/DataFiles/chapter2`, as well as this Amazon S3 location: `s3://packt-serverless-ml-redshift/chapter2/orders.parquet`.

We will use a subset of the `orders` data, which is referenced from the `TPC-H` dataset available here: `https://github.com/awslabs/amazon-redshift-utils/tree/master/src/CloudDataWarehouseBenchmark/Cloud-DWB-Derived-from-TPCH`.

Let's first open the downloaded notebook (`Chapter2.ipynb`) by following these steps:

1. On the Jupyter Notebook landing page, click on **Upload** and open the previously downloaded notebook.

2. Select the kernel (`conda_python3`) once the notebook is uploaded.

> **Note**
> Redshift Serverless requires your `boto3` version to be greater than version 1.24.32.

3. Let's check our `boto3` library version, as shown in *Figure 2.17*.

```
In [1]:  pip show boto3 | grep -i version

         Version: 1.26.35
         Note: you may need to restart the kernel to use updated packages.
```

Figure 2.17 – Checking the boto3 version

If you want to install a specific version greater than 1.24.32, then check the following example:

```
pip install boto3==1.26.35
```

Creating table

As you can see in the `chapter2.ipynb` notebook, we have provided step-by-step instructions to connect to your Redshift Serverless endpoint and perform the necessary operations:

1. Let's start by setting up the parameters and importing the necessary libraries for this exercise. We will set the following two parameters:

    *   `REDSHIFT_WORKGROUP`: The name of the Redshift Serverless workgroup

    *   `S3_DATA_FILE`: The source data file for the load:

        ```
        import boto3
        import time
        import pandas as pd
        import numpy as np
        session = boto3.session.Session()
        region = session.region_name
        REDSHIFT_WORKGROUP = '<workgroup name>'
        S3_DATA_FILE='s3://packt-serverless-ml-redshift/chapter2/orders.
        parquet'
        ```

> **Note**
> Remember to set the parameters as per your settings in the Jupyter notebook.

2. In order to create the table, let's first prepare our DDL and assign it to a `table_ddl` variable:

```
table_ddl = """
DROP TABLE IF EXISTS chapter2.orders CASCADE;
CREATE TABLE chapter2.orders
(o_orderkey       bigint NOT NULL,
 o_custkey        bigint NOT NULL encode az64,
 o_orderstatus    character(1) NOT NULL encode lzo,
 o_totalprice     numeric(12,2) NOT NULL encode az64,
 o_orderdate      date NOT NULL,
 o_orderpriority character(15) NOT NULL encode lzo,
 o_clerk          character(15) NOT NULL encode lzo,
 o_shippriority   integer NOT NULL encode az64,
 o_comment        character varying(79) NOT NULL encode lzo
)
distkey(o_orderkey) compound sortkey(o_orderkey,o_orderdate);"""
```

3. Using the `boto3` library, we will connect to the Redshift Serverless workgroup:

```
client = boto3.client("redshift-data")
```

There are different methods that are available to execute different operations on your Redshift Serverless endpoint. Check out the entire list in this documentation: `https://boto3.amazonaws.com/v1/documentation/api/latest/reference/services/redshift-data.html`.

4. We will use the `execute_statement` method to run an SQL statement, which can be in the **data manipulation language (DML)** or DDL. This method runs a single SQL statement. To run multiple statements, you can use `BatchExecuteStatement`. To get a complete list of different methods and how to use them, please refer to this AWS documentation: `https://docs.aws.amazon.com/redshift-data/latest/APIReference/Welcome.html`:

```
client = boto3.client("redshift-data")
res = client.execute_statement(Database='dev', Sql=table_
ddl,                                WorkgroupName=REDSHIFT_
WORKGROUP)
```

As you can see from the preceding code block, we will first set the client as `redshift-data` and then call `execute_statement` to connect the Serverless endpoint, using the `Database` name and `WorkgroupName`. The method uses temporary credentials to connect to your Serverless workgroup.

We will also pass `table_ddl` as a parameter to create the table. We will create the `Orders` table in our `chapter2` schema.

5. The Redshift Data API sends back a response element once the action is successful, in a JSON format as a dictionary object. One of the response elements is a SQL statement identifier. This value is universally unique and generated by the Amazon Redshift Data API. As you can see in the following code, we have captured the response element, Id, from the output object, res:

```
query_id = res["Id"]
print(query_id)
```

6. In order to make sure that your query is completed, you can use the describe_statement method and pass your id statement as a parameter. This method sends out the response, which contains information that includes when the query started, when it finished, the query status, the number of rows returned, and the SQL statement.

```
##Check status of the Query
status_description = client.describe_statement(Id=query_id)
status = status_description["Status"]
print(status)
```

```
FINISHED
```

Figure 2.18 – Checking the query status

As you can see in *Figure 2.18*, we have captured the status of the statement that we ran, and it sends out the status as FINISHED. This means that we have created our table in the database, and you can verify this by writing a simple SELECT statement against the table.

Loading data using the Redshift Data API

Now, let's move forward to load data into this newly created table. You can use the S3 location for the source data, as mentioned previously. If you use a different S3 location, then remember to replace the path in the parameter (S3_DATA_FILE):

1. Let's write a COPY command, as shown in the following code block. We will create the COPY command in the load_data variable, using the S3 path as a parameter:

```
load_data = f"""COPY chapter2.orders
FROM '{S3_DATA_FILE}'
IAM_ROLE default
FORMAT AS PARQUET;"""
```

2. Next, we will use the execute_statement method to run this COPY command and capture the id statement:

```
res = client.execute_statement(Database='dev', Sql=load_data,
                    WorkgroupName=REDSHIFT WORKGROUP)
query_id = res["Id"]
```

```
print(query_id)
```

Be sure to check whether the status of the query is FINISHED.

3. Once the statement status is defined as FINISHED, we will verify our data load by running a count query, as shown here:

```
cnt = client.execute_statement(Database='dev', Sql='Select
count(1) from chapter2.orders ;', WorkgroupName=REDSHIFT_
WORKGROUP)
query_id = cnt["Id"]
```

We will now print the results:

```
##Print the count query output
results = client.get_statement_result(Id=query_id)
print(results.get('Records'))

[[{'longValue': 1500000}]]
```

Figure 2.19 – Count query results

As you can see in *Figure 2.19*, we have successfully loaded 1.5 million rows.

In the notebook, we have provided a combined code block to show how you can convert all these steps into a function, calling it as and when you require it to load data into a new table.

We also have a GitHub repository (https://github.com/aws-samples/getting-started-with-amazon-redshift-data-api/), which showcases how to get started with the Amazon Redshift Data API in different languages, such as Go, Java, JavaScript, Python, and TypeScript. You can go through the step-by-step process explained in the repository to build your custom application in all these languages, using the Redshift Data API.

Summary

In this chapter, we showcased how you can load data into your Amazon Redshift Serverless database using three different tools and methods, by using the query editor v GUI interface, the Redshift COPY command to load the data, and the Redshift Data API using Python in a Jupyter notebook. All three methods are efficient and easy to use for your different use cases.

We also talked about some of the best practices for the COPY command to make efficient use of it.

In the next chapter, we will start with our first topic concerning Amazon Redshift machine learning, and you will see how you can leverage it in your Amazon Redshift Serverless data warehouse.

3

Applying Machine Learning in Your Data Warehouse

Machine Learning (**ML**) is a routine and necessary part of organizations in today's modern business world. The origins of ML date back to the 1940s when logician Walter Pitts and neuroscientist Warren McCulloch tried to create a neural network that could map out human thought processes.

Organizations can use their data along with ML algorithms to build a mathematical model to make faster, better-informed decisions, and the value of data to organizations today cannot be understated. Data volumes will continue to grow rapidly and organizations that can most effectively manage their data for predictive analytics and identify trends will have a competitive advantage, lower costs, and increased revenue. But to truly unlock this capability, you must bring ML closer to the data, provide self-service tools that do not require a deep data science background and eliminate unnecessary data movement in order to speed up the time it takes to operationalize ML models into your pipelines.

This chapter will introduce you to ML and discuss common use cases to apply ML in your data warehouse. You will begin to see the *art of the possible* and imagine how you can achieve business outcomes faster and more easily through the use of Amazon Redshift ML. We will guide you through the following topics:

- Understanding the basics of ML algorithms
- Traditional steps to implement ML
- Overcoming the challenges of implementing ML
- Exploring the benefits of ML

Understanding the basics of ML

In this section, we will go into more detail about machine learning so that you have a general understanding of the following areas:

- Supervised versus unsupervised learning
- Classification problems
- Regression problems

Let's start by looking at supervised and unsupervised learning.

Comparing supervised and unsupervised learning

A **supervised learning algorithm** is *supervised* by data that contains the known outcome you want to predict. The ML model learns from this known outcome in the data and then uses that learning to predict the outcome of new data.

This known outcome in the data is also referred to as the label or target. For example, if you have a dataset containing home sales information, the sales price would typically be the *target*.

Supervised learning can be further broken down into **classification** or **regression** problems.

With **unsupervised learning** the ML model must learn from the data outcome by grouping data based on similarities, differences, and other patterns without any guidance or known outcome.

You can use unsupervised algorithms to find patterns in the data. For example, you can use unsupervised learning to perform customer segmentation to be more effective in targeting groups of customers. Other use cases include the following:

- Detecting abnormal sensor readings
- Document tagging

With the rich data that data warehouses contain, you can easily get started training models using both supervised and unsupervised learning.

Let's dig into more details on classification and regression problem types.

Classification

Classification problems are tasks to predict class labels, which can be either binary classification or multi-class classification:

- **Binary classification** – The outcome can be in one of two possible classes, for example, to predict whether a customer will churn, whether an email is spam, or whether a patient is likely to be hospitalized after being infected by COVID-19.

- **Multi-class classification** – The outcome can be in one of three or more possible classes – for example, predict a plant species or which category a news article belongs to. Other mutli-class classification use cases include the following:

  - Sales forecasting

  - Intelligent call routing

  - Advertisement optimization

Regression

Regression problems are used when you have a target of continuous values and want to predict a value based on the input variables.

Regression problems are tasks predicting a continuous numeric value:

- **Linear regression**: With linear regression, we predict a numerical outcome such as how much a customer will spend or the predicted revenue for an upcoming concert or sporting event. See *Chapter 7, Building Regression Models*, for more details.

- **Logistic regression**: Logistic regression is another option to solve a binary classification problem. We will show some examples of this technique in *Chapter 6, Building Classification Models*.

Regression use case examples include the following:

- Price and revenue prediction

- Customer lifetime value prediction

- Detecting whether a customer is going to default on a loan

Now we will cover the steps to implement ML.

Traditional steps to implement ML

In this section, you will get a better understanding of the critical steps needed to produce an optimal ML model:

- Data preparation

- Machine learning model evaluation

Data preparation

A typical step in ML is to convert the raw data for input to train your model so that data scientists and data analysts can apply machine learning algorithms to the data. You may also hear the terms **data wrangling** or **feature engineering**.

This step is necessary since machine learning algorithms require inputs to be numbered. For example, you may need outliers or anomalies removed from your data. Also, you may need to fill in missing data values such as missing records for holidays. This helps to increase the accuracy of your model.

Additionally, it is important to ensure your training datasets are unbiased. Machine learning models learn from data and it is important that your training dataset has sufficient representation of demographic groups.

Here are some examples of data preparation steps:

- **Determining the inputs needed for your model** – This is the process of identifying the attributes that most influence the ML model outcome.

- **Cleaning the data** – Correcting data quality errors, eliminating duplicate rows and anomalous data. You need to investigate the data and look for unusual values – this requires knowledge of the domain and how business logic is applied.

- **Transforming the input features** – Machine models require inputs to be numeric. For example, you will use a technique called one-hot encoding when you have data that is not ordinal – such as country or gender data. This will convert the categorical value into a binary value, which creates better classifiers and therefore better models. But as you will see later, when you use the Auto ML feature of Redshift ML, this will have been taken care of for you.

- **Splitting your data into training, validation, and testing datasets:**

  - **Training dataset** – This is a subset of your data that is used to train your model. As a rule of thumb, this is about 80% of your overall dataset.

  - **Validation dataset** – Optionally, you may want to create a validation dataset. This is a ~10% subset of data that is used to evaluate the model during the process of hyperparameter tuning. Examples of hyperparameters include the number of classes (num_class) for multi-class classification and the number of rounds (num_rounds) in an XGBoost model. Note that Amazon Redshift ML automatically tunes your model.

  - **Testing dataset** – This is the remaining 10% of your data used to evaluate the model performance after training and tuning the model.

Traditionally, data preparation is a very time-consuming step and one of the reasons machine learning can be complex. As you will see later, Amazon Redshift ML automates many of the data preparation steps so you can focus on creating your models.

Evaluating an ML model

After you have created your model, you need to calculate the model's accuracy. When using Amazon Redshift ML, you will get a metric to quantify model accuracy.

Here are some common methods used to determine model accuracy:

- **Mean squared error** (MSE): MSE is the average of the squared differences between the predicted and actual values. It is used to measure the effectiveness of regression models. MSE values are always positive: the better a model is at predicting the actual values, the smaller the MSE value is. When the data contains outliers, they tend to dominate the MSE, which might cause subpar prediction performance.

- **Accuracy**: The ratio of the number of correctly classified items to the total number of (correctly and incorrectly) classified items. It is used for binary and multi-class classification. It measures how close the predicted class values are to the actual values. Accuracy values vary between zero and one: one indicates perfect accuracy and zero indicates perfect inaccuracy.

- **F1 score**: The F1 score is the harmonic mean of the precision and recall. It is used for binary classification into classes traditionally referred to as positive and negative. Predictions are said to be true when they match their actual (correct) class and false when they do not. Precision is the ratio of the true positive predictions to all positive predictions (including the false positives) in a dataset and measures the quality of the prediction when it predicts the positive class.

- **F1_Macro** – The F1 macro score applies F1 scoring to multi-class classification. In this context, you have multiple classes to predict. You just calculate the precision and recall for each class, as you did for the positive class in binary classification. F1 macro scores vary between zero and one: one indicates the best possible performance and zero the worst.

- **Area under the curve** (AUC): The AUC metric is used to compare and evaluate binary classification by algorithms such as logistic regression that return probabilities. A threshold is needed to map the probabilities into classifications. The relevant curve is the receiver operating characteristic curve that plots the **true positive rate** (TPR) of predictions (or recall) against the **false positive rate** (FPR) as a function of the threshold value, above which a prediction is considered positive.

Now let's take a look at a couple of these evaluation techniques in more detail.

Regression model evaluation example

A regression model's accuracy is measured by the **Mean Square Error** (**MSE**) and **Root Mean Square Error** (**RMSE**). The MSE is the average squared difference between the predicted values and the actual values in a model's dataset and is also known as *ground truth*. You can square the differences between the actual and predicted answers and then get the average to calculate the MSE. The square root of the MSE computes the RMSE. Low MSE and RMSE scores indicate a good model.

Here is an example of a simple way to calculate the MSE and RMSE so that you can compare them to the MSE score your model generated. Let's assume we have a regression model predicting the number of hotel bookings by a customer for the next month.

Calculate the MSE and RMSE as follows:

```
MSE = (AVG(POWER(( actual_bookings - predicted_bookings)
RMSE = (SQRT(AVG(POWER(( actual_bookings  - predicted_bookings
```

You will calculate the MSE and RMSE for a regression model in one of the exercises in *Chapter 7*.

A classification model can be evaluated based on accuracy. The accuracy method is fairly straightforward, where it can be measured by taking the percentage of the total number of predictions compared to the total number of correct predictions.

Binary classification evaluation example

A confusion matrix is useful for understanding the performance of classification models and is a recommended way to evaluate a classification model. We present the following details for your reference if you want to know more about this topic. We also have a detailed example in *Chapter 10*.

A confusion matrix is in a tabular format and contains four cells – **Actual Values** make up the x axis and **Predicted Values** make up the y axis, and the cells denote **True Positive**, **False Positive**, **False Negative**, and **True Negative**. This is good to measure precision, recall, and the **area under the curve** (**AUC**). *Figure 3.1* shows a simple confusion matrix:

	Actual Values	
	True Positive	False Positive
Predicted Values	False Negative	True Negative

Figure 3.1 – Simple confusion matrix

In *Figure 3.2*, we have 100 records in our dataset for our binary classification model where we are trying to predict customer churn:

	Actual Values	
	10	4
Predicted Values	6	80

Figure 3.2 – Confusion matrix

We can interpret the quality of our predictions from the model as follows:

- Correctly predicted 10 customers would churn

- Correctly predicted 80 customers would not churn

- Incorrectly predicted 4 customers would churn

- Incorrectly predicted 6 customers would not churn

The F1 score is one of the most important evaluation metrics as it considers the precision and recall rate of the model. For example, an F1 score of *.92* means that the model correctly predicted 92% of the time. This method makes sure predictions on both classes are good and *not* biased only toward one class.

Using our confusion matrix example from *Figure 3.2*, we can calculate precision:

$$Precision \ = \ \frac{10}{10 + 4}$$

This could also be written as follows:

$$Precision \ = \ \frac{True\ Positives}{(True\ Positives + False\ Positives)}$$

We can also calculate recall in a similar way:

$$Recall \ = \ \frac{10}{10 + 6}$$

This could also be written as follows:

$$Recall \ = \ \frac{True\ Positives}{(True\ Positives + False\ Negatives)}$$

The F1 score combines precision and recall – it can be calculated as follows:

$$2 \times \left(\frac{precision \times recall}{precision + recall} \right)$$

We have shown you the common techniques for evaluating ML models. As we progress through the book, you will see examples of these techniques that you can apply to your ML evaluation processes.

Now that you have learned the basics of ML, we will discuss some common challenges of implementing ML and how to overcome those.

Overcoming the challenges of implementing ML today

Data growth is both an opportunity and a challenge, and organizations are looking to extract more value from their data. Line-of-business users, data analysts, and developers are being called upon to use this data to deliver business outcomes. These users need easy-to-use tools and don't typically have the skill set of a typical data scientist nor the luxury of time to learn these skills plus being experts in data management. Central IT departments are overwhelmed with analytics and data requirements and are looking for solutions to enable users with self-service tools delivered on top of powerful systems that are easy to use. Following are some of the main challenges:

- Data is more diverse and growing rapidly. We have moved from analyzing terabytes to petabytes and exabytes of data. This data typically is spread across many different data stores across organizations. This means data has to be exported and then landed on another platform to train ML models. Amazon Redshift ML gives you the ability to train models using the data in place without having to move it around.

- A lack of expertise in data management impacts the ability to effectively scale to keep up with volumes of data and an increase in usage.

- A lack of agility to react quickly to events and customer escalations due to data silos and the time required to train a model and make it available for use in making predictions.

- A lack of qualified data scientists to meet today's demands for machine learning. Demands are driven by the need to improve customer experiences, predict future revenues, detect fraud, and provide better patient care, just to name a few.

Consider the following workflow for creating an ML model:

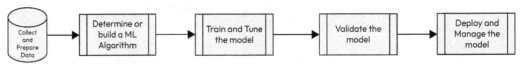

Figure 3.3 – Typical machine learning workflow

Following are the steps to create an ML model, as shown in *Figure 3.3*:

1. First, we start off with data preparation. This can be a very time-consuming process and data may come from many different sources. This data must be cleansed, wrangled, and split into training and test datasets. It then needs to be exported and then loaded into the environment for training.

2. Then you must know which ML algorithm you should use or you need to train your model on. This requires a data scientist who is skilled in tools such as R or Python and has experience in knowing which algorithm is best for a particular problem. As you will see in a later chapter, Amazon Redshift ML can automatically determine the algorithm for you.

3. Then you will iterate many times through training, tuning, and validating the model until you find the best model for your use case.

4. Then, after you deploy the model, you need to continuously monitor the quality of the model and manage the environment including scaling hardware and applying patches and upgrades as needed.

In order to reduce the time required to build data pipelines for machine learning, we must bring machine learning closer to the data and reduce unnecessary data movement. You can use a data architecture, as we talked about in *Chapter 1*, *Introduction to Redshift Serverless*, with the data warehouse at the center. This also includes your data lake and other operational data stores, which, taken together, provide a unified view of all your data that is organized and easily available in a secure manner.

You can build upon the analytic stack that you have built out and enable your data analysts to build and train their own models. All data warehouse users can leverage the power of ML with no data science experience. DW users can create, train, and deploy ML models with familiar SQL commands. Then, using SQL, they can use those models to analyze the data accessible from Amazon Redshift. You can also leverage your existing models in Amazon SageMaker and run inferences on data stores in Amazon Redshift. Data scientists can leverage Redshift ML to iterate faster by baselining models directly through Redshift. BI professionals can now run inference queries directly through tools such as Amazon QuickSight.

Once you implement ML in your organization, you will begin to reap many benefits, which we will explore further in the next section.

Exploring the benefits of ML

There are three main areas where businesses can see the benefits of ML:

*   **Increased revenue** – With ML, you can leverage your data to quickly test new ideas in order to improve customer experiences. For example, using unsupervised learning, you can segment your customers and discover previously unknown purchase patterns, which can drive new focused campaigns for specific product or subscription offerings.

*   **Better operational and financial efficiency** – ML increases automation and agility within your business so that you can respond to changing market conditions faster. One example is forecasting product demand more accurately. By being able to better manage inventory, organizations can see huge cost savings.

*   **Increased agility to respond to business risks** – With ML, you can make decisions quicker than ever before. Using ML to detect anomalies, you can quickly take action when your supply chain, product quality, and other areas of your business face risks.

Application of ML in a data warehouse

Let's look at a few use cases at a high level to illustrate some of these benefits. Subsequent chapters will dive into the details:

- **Improving customer experience**: ML can be used to reduce customer frustration with long wait times. Chatbots can answer many customer questions quickly, and in some cases, all of their questions:

  - **Personalization**: ML can be used to better understand the behaviors and purchase history of customers to make more relevant offerings to customers based on their interests.

  - **Sentiment analysis**: ML can be used to understand customer sentiment from social media platforms. This analysis can then be used for marketing campaigns and customer retention efforts.

- **Predicting equipment maintenance**: Consider any company with a fleet of vehicles or equipment. This could be a package delivery company or a service provider company that must be maintained appropriately. Without ML, it is likely that either equipment will be repaired too soon or too frequently, which leads to higher costs, or equipment will be repaired too late, which leads to equipment being out of service.

  You can use ML to predict the optimal time when each vehicle or piece of equipment needs to have maintenance to maximize operational efficiency.

- **Financial analysis**: Banks and investment companies use ML for automation, risk analysis, portfolio allocation, and much more:

  - **Calculating credit scores** – ML can quickly calculate credit scores and approve loans, which reduces risk.

  - **Fraud detection** – ML can quickly scan large datasets to detect anomalies and flag transactions and automatically decline or approve a transaction. Depending on the nature of a transaction, the system can automatically decline a withdrawal or purchase until a human makes a decision.

- **Sports industry**: Auto racing teams can use a model to predict the best strategies for success and the most effective pit strategy:

  - Build stronger team rosters by predicting future performance

  - Improve player safety by predicting future injuries

- **Health care industry**: Early detection of health conditions by combining ML with historical patient and treatment history and predicting the treatments with the highest probability of success.

These are just some of the benefits of ML. The possibilities are endless and advances are continually being made. As we go through subsequent chapters, you'll see some use cases in action that you can try out on your own and start building up your ML skill set.

Summary

In this chapter, we walked you through how to apply machine learning to a data warehouse and explained the basics of ML. We also discussed how to overcome the challenges of implementing ML so that you can reap the benefits of ML in your organization.

These benefits contribute to increased revenue, better operational efficiencies, and better responses to changing business conditions. After this chapter, you now have a foundational understanding of the use cases and types of models you can deploy in your data warehouse.

In the next chapter, we will introduce you to Amazon Redshift ML and how you can start achieving business outcomes.

Part 2: Getting Started with Redshift ML

Part 2 begins with an overview of Amazon Redshift ML, then dives into how to create various machine learning models using Amazon Redshift ML.

By the end of *Part 2*, you will have an understanding of how to create a model by simply running a SQL command, the difference between supervised and unsupervised learning, and how to solve classification, regression, and clustering problems.

This part comprises the following chapters:

- *Chapter 4, Leveraging Amazon Redshift Machine Learning*
- *Chapter 5, Building Your First Machine Learning Model*
- *Chapter 6, Building Classification Models*
- *Chapter 7, Building Regression Models*
- *Chapter 8, Building Unsupervised Models with K-Means Clustering*

4
Leveraging Amazon Redshift ML

In the previous chapter, we discussed the overall benefits of **machine learning** (**ML**) and how it fits into your data warehouse.

In this chapter, we will focus specifically on how to leverage **Amazon Redshift ML** to solve various use cases. These examples are designed to give you the foundation you need as you get hands-on training models, beginning in *Chapter 5*. We will show the benefits of Redshift ML, such as eliminating data movement, being able to create models using simple SQL, and drastically reducing the time it takes to train a new model and make it available for inference. Additionally, you will learn how Amazon Redshift ML leverages **Amazon SageMaker** behind the scenes to automatically train your models as we guide you through the following main topics:

- Why Amazon Redshift ML?
- An introduction to Amazon Redshift ML
- A CREATE MODEL overview

Why Amazon Redshift ML?

Amazon Redshift ML gives you the ability to create and train ML models with simple SQL commands, without the need to build specialized skills. This means your data analysts, data engineers, and BI analysts can now leverage their SQL skills to do ML, which increases agility, since they no longer need to wait for an ML expert to train their model.

Additionally, since you use your model in the data warehouse, you no longer need to export data to be trained or import it back into the warehouse after your model is used to make predictions.

You do not have to worry about managing the governance of data. Data never leaves your VPC when you export data for training.

You can control who can create models and who can run inference queries on those models.

Amazon Redshift ML provides a very cost-effective solution for training and using models. The cost for Amazon SageMaker resources is based on the number of cells in your training dataset, which is the product of the number of rows times the number of columns in the training set.

The costs for running prediction queries using Amazon Redshift Serverless are based on the compute capacity used by your queries.

To learn more about Amazon Redshift Serverless costs refer here `https://docs.aws.amazon.com/redshift/latest/mgmt/serverless-billing.html`.

You have the ability to control the costs of model training by limiting how much data is used to train the model, and by controlling the time for training. We will show you examples of this later in the *A CREATE MODEL overview* section.

When you run a prediction query, all predictions are computed locally in your Redshift data warehouse. This enables you to achieve very high throughput and low latency.

An introduction to Amazon Redshift ML

By leveraging Amazon Redshift ML, your organization can achieve many benefits. First of all, you eliminate unnecessary data movement, users can use familiar SQL commands, and integration with Amazon SageMaker is transparent.

Let's define some of the terms that you will see throughout the remaining chapters:

- **CREATE MODEL**: This is a command that will contain the SQL that will export data to be used to train your model.

- **Features**: These are the attributes in your dataset that will be used as input to train your model.

- **Target**: This is the attribute in your dataset that you want to predict. This is also sometimes referred to as a **label**.

- **Inference**: This is also referred to as **prediction**. In Amazon Redshift ML, this is the process of executing a query against a trained model to get the predicted value generated by your model.

To be able to create and access your ML models in Amazon Redshift to run prediction queries, you need to grant permissions on the model object, just like you would on other database objects such as tables, views, or functions.

Let's assume you have created the following role to allow a set of users to create models, called `analyst_cm_role`. A superuser can grant permissions to this role as follows:

```
GRANT CREATE MODEL to role analyst_cm_role
```

Users/groups/roles with the `CREATE MODEL` privilege can create a model in any schema in your serverless endpoint or Redshift cluster if the user has the `CREATE` permission on the Schema. A Redshift ML model is part of the schema hierarchy, similar to tables, views, stored procedures, and user-defined functions. Let's assume we have a schema called `demo_ml`. You can grant `CREATE` and `USAGE` privileges on the `demo_ml` schema to the analyst role using the following `GRANT` statement:

```
GRANT CREATE, USAGE ON SCHEMA demo_ml TO role analyst_cm_role
```

Now, let's assume we have another role to allow a set of users access to run prediction queries called `analyst_prediction_role`. You can grant access to run predictions on models using the following:

```
GRANT EXECUTE ON MODEL demo_ml.customer_churn_auto_model TO role
analyts_prediction_role
```

The source data to create a model can be in Redshift or any other source that you can access from Redshift, including your **Amazon Simple Storage Service (Amazon S3)** S3 data lake via Spectrum or other sources using the Redshift federated query capability. At the time of writing, Amazon Aurora and Amazon RDS for PostgreSQL and MySQL are supported. More details are available here: `https://docs.aws.amazon.com/redshift/latest/dg/federated-overview.html`.

Amazon Redshift ML and Amazon SageMaker manage all data conversions, permissions, and resource usage. The trained model is then compiled by SageMaker Neo and made available as a user-defined function in Amazon Redshift so that users can make predictions using simple SQL.

Once your model is trained and available as a function in Amazon Redshift, you can run prediction queries at scale and efficiently, locally in Amazon Redshift.

See the process flow here in *Figure 4.1*:

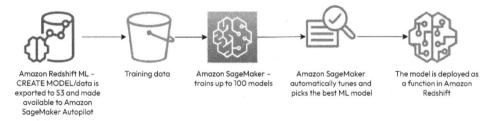

Figure 4.1 – The Redshift ML CREATE MODEL process flow

Now, let us go into more detail on how you can use the `CREATE MODEL` statement.

A CREATE MODEL overview

The CREATE MODEL statement allows for flexibility when addressing the various use cases you may need. There are four main types of CREATE MODEL statements:

- AUTO everything
- AUTO with user guidance, where a user can provide the problem type
- AUTO OFF, with customized options provided by the user
- **Bring your own model (BYOM)**

Figure 4.2 illustrates the flexibility available when training models with Amazon Redshift ML:

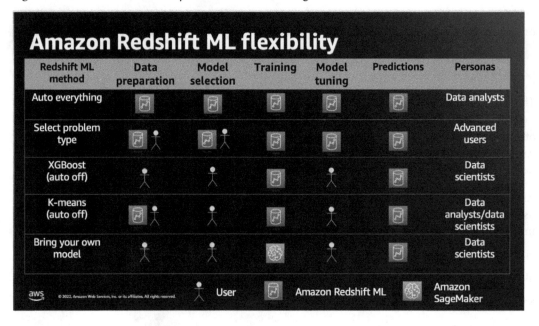

Figure 4.2 – Amazon Redshift ML flexibility

In this chapter, we will provide an overview of the various types of CREATE MODEL statements. Subsequent chapters will provide in-depth examples of how to create all the different types of models, load the data to Redshift, and split your data into training and testing datasets.

In this section, we will walk you through the options available to create models and the optional parameters available that you can specify. All of the examples in this chapter are informational to prepare you for the remaining chapters. You will create your first model in *Chapter 5*.

AUTO everything

When you execute a CREATE MODEL command to solve a supervised learning problem using AUTO everything, Amazon Redshift ML and Amazon SageMaker manage all the data preprocessing, model training, and model tuning for you. Data will be exported from Amazon Redshift to Amazon S3, where SageMaker will train and tune up to 100 models. **SageMaker Autopilot** will automatically determine the algorithm and problem type. The best-trained model is then compiled by SageMaker Neo and made available as a user-defined function in Amazon Redshift so that users can make predictions using simple SQL.

See the following syntax for an AUTO everything model:

```
CREATE MODEL model_name
    FROM { table_name | ( select_query ) }
    TARGET column_name
    FUNCTION prediction_function_name
    IAM_ROLE { default }
    SETTINGS (
      S3_BUCKET 'bucket',
      [ MAX_CELLS integer ]
    )
```

You simply supply a table name or SQL statement for the data you want to use in training, along with the TARGET column that you are trying to predict.

Let's apply this to a simple example. Let's assume we have a table called reservation_history that contains hotel reservation data, and we want to determine whether guests are likely to cancel an upcoming reservation:

```
CREATE TABLE reservation_history (
customerid bigint ,
city character varying(50),
reservation_date timestamp without time zone,
loyalty_program character (1),
age bigint,
marital_status character (1),
cancelled character (1)
)
DISTSTYLE AUTO;
```

The CREATE MODEL statement would look like this (note that this is informational; you do not need to run this):

```
CREATE MODEL predict_guest_cancellation
    FROM reservation_history
    TARGET cancelled
    FUNCTION predict_cancelled_reservation
    IAM_ROLE default
    SETTINGS (
      S3_BUCKET '<<your-s3-bucket>>'
)
```

In this CREATE MODEL statement, we only provided the minimum required parameters, which are IAM_ROLE and S3_BUCKET. The TARGET parameter is cancelled, which is what we will try to predict, based on the input we send to the CREATE MODEL statement. In this example, we send everything from the reservation_history table. The FUNCTION name is a description of the function that will be used later for predictions. The IAM_ROLE parameter will be attached to your serverless endpoint and provides access to SageMaker and an **S3** bucket, which will contain the artifacts generated by your CREATE MODEL statement. Refer to *Chapter 2*, where we showed how to set up an IAM role.

Amazon SageMaker will automatically determine that this is a binary classification model, since our TARGET can only be one of two possible values. Amazon SageMaker will also choose the best model type. At the time of writing, the supported model types for supervised learning are as follows:

- XGBoost: Based on the gradient-boosted trees algorithm
- Linear Learner: Provides an increase in speed to solve either classification or regression problems
- MLP: A deep learning algorithm using a multilayer perceptron

You will create models using each of these models in subsequent chapters.

AUTO with user guidance

More advanced users with a good understanding of ML may wish to provide more inputs to a model, such as model _type, problem_type, preprocesors, and objective.

Using our reservation example, we will build on the AUTO capabilities and specify a few more parameters:

- MODEL_TYPE: XGBoost
- PROBLEM_TYPE: binary_classification
- Objective: F1

- S3_GARBAGE_COLLECT – OFF: If set to OFF, the resulting datasets used to train the models remain in Amazon S3 and can be used for other purposes, such as troubleshooting

- MAX_RUNTIME – 1800: This is one way to control the costs of model training by limiting the training time to 1800 seconds; the default is 5400 seconds

By specifying MODEL_TYPE and/or PROBLEM_TYPE along with the Objective parameters, you can shorten the amount of time needed to train a model, since SageMaker does not have to determine these. Here is an example of the CREATE MODEL statement:

```
CREATE MODEL predict_guest_cancellation
    FROM reservation_history
    TARGET cancelled
    FUNCTION predict_cancelled_reservation
    IAM_ROLE default
    MODEL_TYPE XGBoost
    PROBLEM_TYPE BINARY_CLASSIFICATION
    OBJECTIVE 'F1'
    SETTINGS (
      S3_BUCKET '<<your-S3-bucket>>',
      MAX_RUNTIME 1800
);
```

> **Note**
>
> Increasing MAX_RUNTIME and MAX_CELLS often improves model quality by allowing SageMaker to explore more candidates. If you want faster iteration or exploration of your dataset, reduce MAX_RUNTIME and MAX_CELLS. If you want improved accuracy of models, increase MAX_RUNTIME and MAX_CELLS.

It is a good practice to specify the problem type and objective, if known, to shorten training time. To improve model accuracy, provide more data if possible and include any features (input) that can influence the target variable.

Additionally, you can add your own preprocessors by specifying transformers. At the time of writing, Amazon Redshift ML supports 10 transformers including OneHotEncoder, Ordinal Encoder, and StandardScaler. You can find the complete list here: https://docs.aws.amazon.com/redshift/latest/dg/r_create_model_use_cases.html#r_user_guidance_create_model.

Amazon Redshift ML stores the trained transformers and automatically applies them as part of the prediction query. You don't need to specify them when generating predictions from your model.

Let's take, as an example, using OneHotEncoder, which is used to convert a categorical value such as country or gender into a numeric value (binary vector) so that ML algorithms can better do predictions. Let's create a model using one-hot encoding for our input columns, marital_status and loyalty_program. Note that this model is an example, and you do not need to run this statement:

```
CREATE MODEL predict_guest_cancellation
    FROM reservation_history
    TARGET cancelled
    FUNCTION predict_cancelled_reservation
    IAM_ROLE default
    MODEL_TYPE XGBoost
    PROBLEM_TYPE BINARY CLASSIFICATION
    OBJECTIVE 'F1'
    PREPROCESSORS '[
      {"ColumnSet": [
        "loyalty_program",
        "marital_status"
        ],
        "Transformers" :[
          "OneHotEncoder"
          ]
        ]
        }
    ]'

    SETTINGS (
    S3_BUCKET '<<your-S3-bucket>>',
    MAX_RUNTIME 1800
);
```

So far, all the CREATE MODEL examples we showed use AUTO ON. This is the default if you do not specify this parameter. Now, let's move on to how you can do your own model tuning using AUTO OFF with XGBoost.

XGBoost (AUTO OFF)

As an ML expert, you have the option to do hyperparameter tuning by using the AUTO OFF option with the CREATE MODEL statement. This gives you full control and Amazon Redshift ML does not attempt to discover the optimal preprocessors, algorithms, and hyperparameters.

Let's see what the CREATE MODEL syntax looks like using our example reservation dataset.

We will specify the following parameters:

- AUTO OFF: Turns off the automatic discovery of a preprocessor, an algorithm, and hyperparameters

- MODEL_TYPE:- xgboost

- OBJECTIVE: 'binary:logistic'

- PREPROCESSORS: 'none'

- HYPERPARAMETERS: DEFAULT EXCEPT(NUM_ROUND '100'/)

Refer here for a list of hyperparameters for XGBoost: https://docs.amazonaws.cn/en_us/redshift/latest/dg/:

```
r_create_model_use_cases.html#r_auto_off_create_model
```

As of this writing, 'none' is the only available option to specify for PREPROCESSORS when using AUTO OFF. Since we cannot specify one-hot encoding, we can use a case statement with our SQL to apply this:

```
CREATE MODEL predict guest_cancellation
    FROM
        (Select customerid,
      city,
      reservation_date,
      case when loyalty_program = 'Y' then 1 else 0 end as loyalty_
program_y,
      case when loyalty_program = 'N' then 1 else 0 end as loyalty_
program_n,
      age,
      case when marital_status = 'Y' then 1 else 0 end as married,
      case when marital_status = 'N' then 1 else 0 end as not_
married,
      cancelled
      from reservation_hitory)

    TARGET cancelled
    FUNCTION predict_cancelled_reservation
```

```
      IAM_ROLE default
      AUTO OFF
      MODEL_TYPE XGBoost
      OBJECTIVE 'binary:logistic'
      PREPROCESSORS 'none'
      HYPERPARAMETERS DEFAULT EXCEPT (NUM_ROUND '100')
      SETTINGS (
        S3_BUCKET 'bucket',
        MAX_RUNTIME 1800
  );
```

In *Chapter 10*, you will build an XGBoost model using AUTO OFF and gain a better understanding of this option.

Now, let's take a look at another AUTO OFF option using the K-means algorithm.

K-means (AUTO OFF)

The K-means algorithm is used to group data together that isn't labeled. Since this algorithm discovers groupings in your data, it solves an "*unsupervised*" learning problem.

Let's see what a sample CREATE MODEL looks like if we want to group our reservation_ history data:

- AUTO OFF: Turns off the automatic discovery of a preprocessor, an algorithm, and hyperparameters

- MODEL_TYPE: KMEANS

- PREPROCESSORS: OPTIONAL (at the time of writing, Amazon Redshift supports StandScaler, MinMax, and NumericPassthrough for KMEANS)

- HYPERPARAMETERS: DEFAULT EXCEPT (K 'N'), where N is the number of clusters you want to create

Here is an example of a CREATE MODEL statement. Note that you will not run this statement:

```
CREATE MODEL guest_clusters
    FROM
        (Select
        city,
        reservation_date,
        loyalty_program,
        age,
        marital_status
        from reservation_hitory)
```

```
      FUNCTION get_guest_clusters
      IAM_ROLE default
      AUTO OFF
      MODEL_TYPE KMEANS
      PREPROCESSORS 'none'
      HYPERPARAMETERS DEFAULT
      EXCEPT (K '5')
      SETTINGS (
         S3_BUCKET '<<your-S3-bucket>>'
  );
```

Note that we are creating five clusters with this model. With the K-means algorithm, it is important to experiment with a different number of clusters. In *Chapter 8*, you will get to dive deep into creating K-means models and determining how to validate the optimal clusters.

Now, let's take a look at how you can run prediction queries using models built outside of Amazon Redshift ML.

BYOM

Additionally, you can use a model trained outside of Amazon Redshift with Amazon SageMaker for either local or remote inference in Amazon Redshift.

Local inference

Local inference is used when models are trained outside of Redshift in Amazon SageMaker. This allows you to run inference queries inside of Amazon Redshift without having to retrain a model.

Let's suppose our previous example of building a model to predict whether a customer will cancel a reservation was trained outside of Amazon Redshift. We can bring that model to Redshift and then run inference queries.

Our CREATE MODEL sample will look like this:

- model_name: This is the name you wish to give the local model in Redshift
- FROM: This is job_name from Amazon SageMaker – you can find this in Amazon SageMaker under **Training Jobs**
- FUNCTION: The name of the function to be created along with the input data types

- RETURNS: The data type of the value returned by the function:

```
CREATE MODEL predict_guest_cancellation_local_inf
    FROM 'sagemaker_job_name'
    FUNCTION predict_cancelled_reservation_local(bigint,
varchar, timestamp, char, bigint, char)
    RETURNS char
    IAM_ROLE default
    SETTINGS (
      S3_BUCKET '<<your-S3-bucket>>' );
```

Note that the data types in `FUNCTION` match the data types from our `reservation_history` table, and `RETURNS` matches the data type of our `TARGET` variable, which is `cancelled`.

You can derive the SageMaker `JobName` by navigating to the AWS Management Console and going to SageMaker:

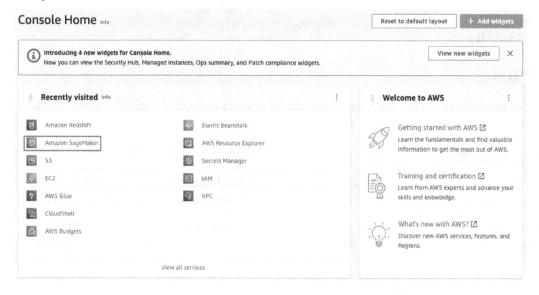

Figure 4.3 – Console Home

After clicking on **Amazon SageMaker**, click on **Training jobs**, as shown in *Figure 4.4*:

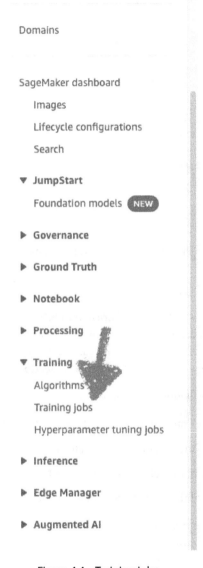

Figure 4.4 – Training jobs

Next, note the job name of the model you wish to use for local inference, which is what you will put in your CREATE MODEL statement (see *Figure 4.5*):

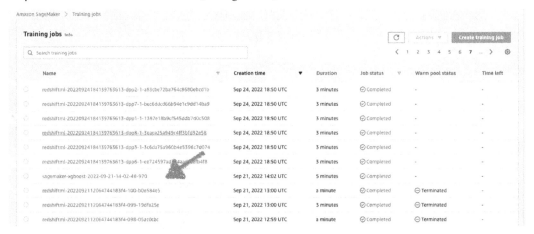

Figure 4.5 – The training job name

Remote inference

Remote inference is useful if you have a model created in SageMaker for an algorithm that is not available natively in Amazon Redshift ML. For example, anomaly detection can be done using the Random Cut Forest algorithm from SageMaker. You can create a model that references the endpoint of the SageMaker model and then be able to run anomaly detection in Amazon Redshift.

Our CREATE MODEL sample will look like this:

- model_name: The name you wish to give the local model in Redshift
- FUNCTION: The name of the function to be created along with the input data types
- RETURNS: The data type of the value returned by the function
- SAGEMAKER: The name of the Amazon SageMaker endpoint:

```
CREATE MODEL random_cut_forest
FUNCTION remote_fn_rcf(int)
RETURNS decimal(10,6)
SAGEMAKER 'sagemaker_endpoint'
IAM_ROLE  default;
```

Note that the data types in FUNCTION are for the input we send, and RETURNS is the data type of the data we receive when invoking the function.

You can derive the SageMaker endpoint by navigating to the AWS Management Console, going to SageMaker, and then clicking on **Endpoints**:

Search

▼ **JumpStart**

Foundation models

▶ **Governance**

▶ **Ground Truth**

▶ **Notebook**

▶ **Processing**

▼ **Training**

Algorithms

Training jobs

Hyperparameter tuning jobs

▼ **Inference**

Marketplace model
packages

Models

Endpoint configurations

Endpoints

Batch transform jobs

Shadow tests

▶ **Edge Manager**

▶ **Augmented AI**

▶ **AWS Marketplace**

Figure 4.6 – Endpoints

After you click on **Endpoints**, as shown in *Figure 4.6*, you can see the endpoint names, as shown in *Figure 4.7*:

Figure 4.7 – The endpoint names

Then, note the name of the endpoint for the model you wish to use for remote inference and put it in your CREATE MODEL statement.

You will dive deep into BYOM in *Chapter 11* and get hands-on experience creating models for both local and remote inference.

Summary

In this chapter, we discussed why Amazon Redshift ML is a good choice to use data in your data warehouse to make predictions.

By bringing ML to your data warehouse, Amazon Redshift ML enables you to greatly shorten the amount of time to create and train models by putting the power of ML directly in the hands of your developers, data analysts, and BI professionals.

Your data remains secure; it never leaves your VPC. Plus, you can easily control access to create and use models.

Finally, we showed you different methods of creating models in Redshift ML, such as using AUTO, how to guide model training, and an advanced method to supply hyperparameters.

Now, you understand how ML fits into your data warehouse, how to use proper security and configuration guidelines with Redshift ML, and how a model is trained in Amazon SageMaker.

In the next chapter, you will get hands-on and create your first model using Amazon Redshift ML, learn how to validate the model, and learn how to run an inference query.

5

Building Your First Machine Learning Model

In the previous chapter, you learned about **Redshift Machine Learning** (**ML**) benefits such as eliminating data movement and how models can be created using simple **Structured Query Language** (**SQL**) commands.

In this chapter, you are going to build your first machine learning model by using the standard SQL dialect. Amazon Redshift makes it very easy to use familiar SQL dialect to train, deploy, and run inferences against machine learning models. This approach makes it easy for different data personas, for example, database developers, database engineers, and citizen data scientists, to train and build machine learning models without moving data outside of their data warehouse platform and without having to learn a new programming language.

In this chapter, you will learn about using Amazon Redshift ML simple CREATE MODEL, which uses the **Amazon SageMaker Autopilot** framework behind the scenes, to create your first model. You will also learn how to evaluate a model to make sure the model performance is good and that it is usable and not biased. When you are done with this chapter, you should be familiar with the Redshift ML simple CREATE MODEL command and different methods used to evaluate your ML model.

In this chapter, to build your first machine learning model, we will go through the following main topics:

- Redshift ML simple CREATE MODEL
- Evaluating model performance

Technical requirements

This chapter requires a web browser and the following:

- An AWS account.
- An Amazon Redshift Serverless endpoint.

- Amazon Redshift Query Editor v2.

- Completing the *Getting started with Amazon Redshift Serverless* section in *Chapter 1*.

You can find the code used in this chapter here: `https://github.com/PacktPublishing/Serverless-Machine-Learning-with-Amazon-Redshift/`

Data files required for this chapter are located in a public S3 bucket: `s3://packt-serverless-ml-redshift/`

Let's begin!

Redshift ML simple CREATE MODEL

Redshift ML simple CREATE MODEL is a feature in Amazon Redshift that allows users to create machine learning models using SQL commands, without the need for specialized skills or software. It simplifies the process of creating and deploying machine learning models by allowing users to use familiar SQL syntax to define the model structure and input data, and then automatically generates and trains the model using Amazon SageMaker. This feature can be used for a variety of machine learning tasks, including regression, classification, and clustering.

Before we dive into building the first ML model, let us set the stage by defining a problem statement that will form the basis of our model-building solution.

We are going to use a customer sales dataset to build the first machine learning model. Business leaders at the fictitious *ABC Company* are grappling with dwindling sales. The data team at *ABC Company* has performed descriptive and diagnostic analytics and determined that the cause of decreasing sales is departing customers. To stop this problem, data analysts who are familiar with SQL language and some machine learning concepts have tapped into Redshift ML. Business users have documented which customers have and have not churned and teamed up with data analysts.

To solve the business problem, the data analysts start by analyzing the sales dataset. With Redshift SQL commands, they will write SQL aggregate queries and create visualizations to understand the trends. The data analyst team then creates an ML model using the Redshift ML simple CREATE MODEL command. Finally, the data analysts evaluate the model performance to make sure the model is useful.

Uploading and analyzing the data

The dataset used for this chapter is located here: `s3://packt-serverless-ml-redshift/`. We have modified the dataset to better fit the chapter's requirements.

Dataset citation

This dataset is attributed to the University of California Irvine Repository of Machine Learning Datasets (Jafari-Marandi, R., Denton, J., Idris, A., Smith, B. K., & Keramati, A. (2020). *Optimum Profit-Driven Churn Decision Making: Innovative Artificial Neural Networks in Telecom Industry. Neural Computing and Applications.*

This dataset contains customer churn information. The following table lists the metadata of the dataset:

Name	Data Type	Definition
state	varchar(2)	US state in which the customer is located
account_length	int	Length of customer account
area_code	int	Area code or zip code of the customer
phone	varchar(8)	Phone number of the customer
intl_plan	varchar(3)	International plan subscriber
vMail_plan	varchar(3)	Voicemail plan subscriber
vMail_message	int	Voicemail message subscriber
day_mins	float	Aggregated daily minutes
day_calls	int	Aggregated daily calls
day_charge	float	Aggregated daily charges
total_charge	float	Total charges
eve_mins	float	Evening minutes
eve_calls	int	Evening calls
eve_charge	float	Evening charges
night_mins	float	Nightly minutes
night_calls	int	Nightly calls
night_charge	float	Nightly charges
intl_mins	float	International minutes
intl_calls	int	International calls
intl_charge	float	International charges
cust_serv_calls	int	Number of calls to customer service
churn	varchar(6)	Whether customer churned or not
record_date	date	Record updated date

Table 5.1 – Customer call data

After successfully connecting to Redshift as an admin or database developer, create the schema and load data into Amazon Redshift as follows:

1. Navigate to **Redshift query editor v2**, connect to the **Serverless:default** endpoint, and connect to the **dev** database.

    Create a new editor and rename the `untitled` query editor by saving it as `Chapter5`, as shown in *Figure 5.1*:

Figure 5.1 – Connecting to query editor v2

2. Create a Redshift schema named `Chapter5_buildfirstmodel`. Redshift schemas contain tables, views, and other named objects. For this chapter, tables and machine learning models will be created in this schema:

    ```
    Create schema chapter5_buildfirstmodel;
    ```

3. Create a Redshift table named `customer_calls_fact`. This table is used to load the dataset that has customer call information. This table is natively created in Redshift and used for training and validating the Redshift ML model:

    ```
    CREATE TABLE IF NOT EXISTS chapter5_buildfirstmodel.customer_
    calls_fact (
    state varchar(2),
    account_length int,
    area_code int,
    phone varchar(8),
    intl_plan varchar(3),
    vMail_plan varchar(3),
    vMail_message int,
    day_mins float,
    day_calls int,
    day_charge float,
    total_charge float,
    eve_mins float,
    ```

```
eve_calls int,
eve_charge float,
night_mins float,
night_calls int,
night_charge float,
intl_mins float,
intl_calls int,
intl_charge float,
cust_serv_calls int,
churn varchar(6),
record_date date)
Diststyle AUTO;
```

4. Load the customer call data into the Redshift table by using the following command:

```
COPY  chapter5_buildfirstmodel.customer_calls_fact
FROM 's3://packt-serverless-ml-redshift/chapter05/customerdime/'
IAM_ROLE default
delimiter ',' IGNOREHEADER 1
region 'eu-west-1';
```

We use the Redshift COPY command to load the data into our table. COPY commands load data in parallel into a Redshift table. You can load terabytes of data by using the COPY command.

5. In the final step, we will analyze the customer churn fact table by creating a histogram for customer churn. To do this, let's use the query editor v2 chart feature to create a histogram chart. In order to create the histogram, we need to count the number of customers who have churned and not churned. To get this information, first, run the following command:

```
SELECT churn, count(*) Customer_Count FROM chapter5_
buildfirstmodel.customer_calls_fact
GROUP BY churn
;
```

Now, click on the **Chart** option found on the right-hand side in the **Result** pane to view the histogram:

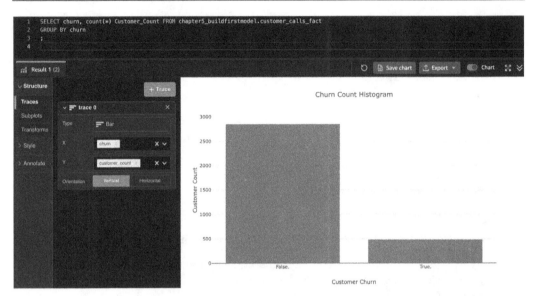

Figure 5.2 – Customers churned versus not churned histogram

From the preceding chart, you can see that the `customer_calls_fact` table has **3333** customers, of which **483** have churned.

Now, we analyzed the dataset and found that there are customers who have churned. The next step is to create a machine learning model. For this, we will use the Redshift ML simple `CREATE MODEL` method.

Diving deep into the Redshift ML CREATE MODEL syntax

Since this is the first time you are going to use the `CREATE MODEL` syntax, let's refresh the basic constructs of the command here.

Redshift ML provides the easy-to-use `CREATE MODEL` syntax to create ML models. In this section, we will focus on a simple form of the `CREATE MODEL` command. In later chapters, you will learn about other forms of creating model statements.

Simple `CREATE MODEL` is the most basic form of Redshift `CREATE MODEL` statement. It is geared toward the personas who are not yet ready to deal with all the intricacies of the machine learning process. This form of model creation is also used by experienced personas such as citizen data scientists for its simplicity in creating a machine learning model. Data cleaning is an essential step for any ML problem, otherwise, it follows the principle of *garbage in, garbage out*. Data cleaning still remains a necessary task, however, with Redshift ML data transformation, standardization and model selection won't be necessary.

We use the following command for simple model creation:

```
CREATE MODEL model_name
    FROM { table_name | ( select_query ) }
    TARGET column_name
    FUNCTION prediction_function_name
    IAM_ROLE { default }
    SETTINGS (
      S3_BUCKET 'bucket',
      [ MAX_CELLS integer ]
    )
```

In the preceding CREATE MODEL syntax, as a user, you specify your dataset – in our case, customer_ calls_fact – in the FROM clause. We set the variable that we are targeting to predict, in our case churn, in the TARGET parameter. As a user, you also give a name to the function, which you will use in select queries to run predictions.

For more information about simple CREATE MODEL parameters, please refer to the Redshift public document here: https://docs.aws.amazon.com/redshift/latest/dg/r_create_ model_use_cases.html#r_simple_create_model

We've learned about the generic simple CREATE MODEL syntax. Now, let's create the syntax for our dataset and run it.

Creating your first machine learning model

Finally, we will now build our first ML model to predict customer churn events. As this is our first machine learning model, let's use the simple CREATE MODEL command. This option uses Amazon SageMaker Autopilot, which means, without the heavy lifting of building ML models, you simply provide a tabular dataset and select the target column to predict and SageMaker Autopilot automatically explores different solutions to find the best model. This includes data preprocessing, model training, and model selection and deployment. AutoMode is the default mode:

1. Redshift ML shares training data and artifacts between Amazon Redshift and SageMaker through an S3 bucket. If you don't have one already, you will need to create an S3 bucket. To do this, navigate to the Amazon S3 console and click on the **Create bucket** button:

Figure 5.3 – S3 console

2. On the **Create bucket** page, under **Bucket name**, provide a name, for example, `serverless machinelearningwithredshift-<your account id>`, where `<your account id>` is your AWS account number.

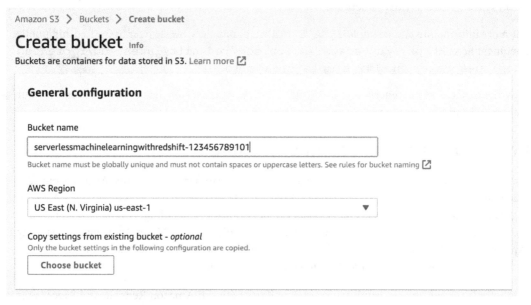

Figure 5.4 – Creating an S3 bucket

3. Before we send our dataset to the `CREATE MODEL` command, we will split the dataset into two parts – one is the training dataset, which is used to train the machine learning model, and the other one is for testing the model once it is created. We do this by filtering customer records that have `record_date` of less than `'2020-08-01'` for training and `record_date` greater than `'2020-07-31'` for testing. Run the following queries to check our record split:

```
select sum(case when record_date <'2020-08-01' then 1 else 0
end) as Training_Data_Set,
```

```
sum(case when record_date >'2020-07-31' then 1 else 0 end) as
Test_Data_Set
from chapter5_buildfirstmodel.customer_calls_fact
```

In *Figure 5.5*, we can see we have **2714** records in the training set and **619** records in the test set.

Figure 5.5 – Training and test dataset record count

We apply the filtering condition when training and testing the model on our dataset. In the next step, we are going to create the model using this filter condition on our dataset.

4. Now run the following code to create `customer_churn_model`. Make sure to replace `<your account id>` with the correct AWS account number. Please note that since we are going to use simple `CREATE MODEL`, we set the max allowed time through the `MAX_RUNTIME` parameter. This is the maximum training time that Autopilot will take. We have set it to 1,800 seconds, which is 30 minutes. If you don't specify a value for `MAX_RUNTIME` it will use the default value of 5,400 seconds (90 minutes):

```
CREATE MODEL chapter5_buildfirstmodel.customer_churn_model
FROM (SELECT state,
             account_length,
             area_code,
             phone,
             intl_plan,
             vMail_plan,
             vMail_message,
             day_mins,
             day_calls,
             day_charge,
             total_charge,
             eve_mins,
             eve_calls,
             eve_charge,
             night_mins,
             night_calls,
             night_charge,
             intl_mins,
             intl_calls,
             intl_charge,
             cust_serv_calls,
```

```
                    replace(churn,'.','') as churn
          FROM chapter5_buildfirstmodel.customer_calls_fact
            WHERE record_date < '2020-08-01'


        )
TARGET churn
FUNCTION predict_customer_churn
IAM_ROLE default
SETTINGS (
  S3_BUCKET 'serverlessmachinelearningwithredshift-<your account
id>',
  MAX_RUNTIME 1800
)
;
```

Let us understand more about the preceding command:

- The `SELECT` query in the `FROM` clause specifies the training data

- The `TARGET` clause specifies which column is the label for which the `CREATE MODEL` statement builds a model to predict

- The other columns in the training query are the features (input) used to predict the churn variable

- The `predict_customer_churn` function is the name of an inference function used in `SELECT` queries to generate predictions

- `S3_Bucket` is the location where Redshift ML saves artifacts when working with SageMaker

- Having `MAX_RUNTIME` set as 1,800 seconds specifies the maximum time that SageMaker will take to train our model

After you run the `CREATE MODEL` command, run the following command to check the status of the model:

```
SHOW MODEL chapter5_buildfirstmodel.customer_churn_model;
```

The Redshift ML `CREATE MODEL` statement is asynchronous, which means that when the model is under training, the query shows it is completed and the training is happening in Amazon SageMaker. To find out the status of the model, run the `SHOW MODEL` command.

In the following screenshot, you can see the SHOW MODEL output shows **Model State** as **TRAINING**:

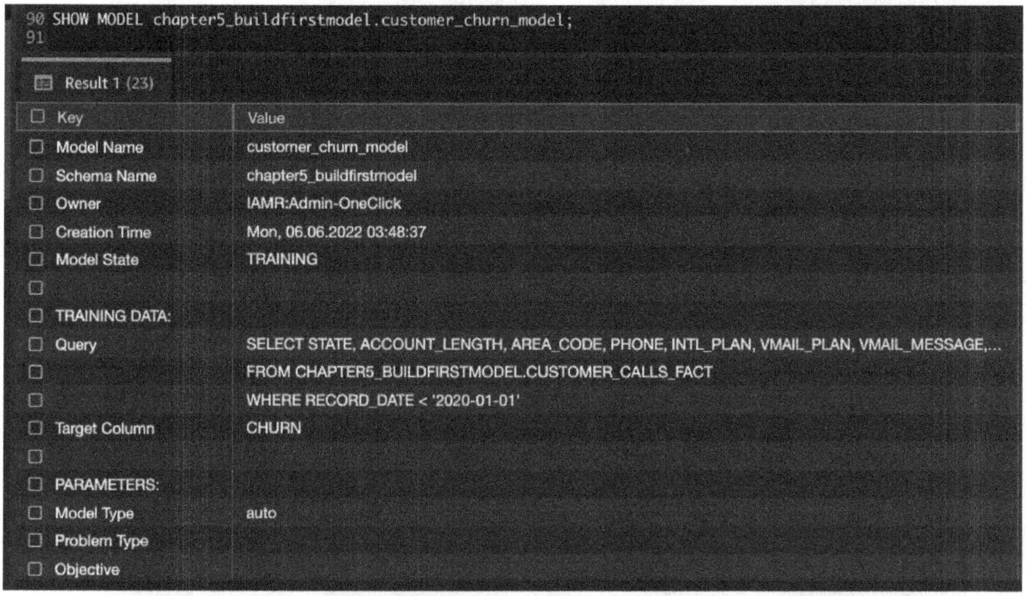

```
90  SHOW MODEL chapter5_buildfirstmodel.customer_churn_model;
91
```

Key	Value
Model Name	customer_churn_model
Schema Name	chapter5_buildfirstmodel
Owner	IAMR:Admin-OneClick
Creation Time	Mon, 06.06.2022 03:48:37
Model State	TRAINING
TRAINING DATA:	
Query	SELECT STATE, ACCOUNT_LENGTH, AREA_CODE, PHONE, INTL_PLAN, VMAIL_PLAN, VMAIL_MESSAGE,...
	FROM CHAPTER5_BUILDFIRSTMODEL.CUSTOMER_CALLS_FACT
	WHERE RECORD_DATE < '2020-01-01'
Target Column	CHURN
PARAMETERS:	
Model Type	auto
Problem Type	
Objective	

Figure 5.6 – Model State TRAINING

When the same SHOW MODEL command is run after a while, **Model State** is displayed as **READY**, which means data processing, model training, model selection, and model deployment to Redshift is completed successfully. From the following screenshot, you can see that **Model Status** now shows **READY**. You can also see the **Estimated Cost** value, which represents Amazon SageMaker training hours. This value does not equal the elapsed training time as it is an accumulation of training time on the SageMaker instances used.

Key	Value
☐ Model Name	customer_churn_model
☐ Schema Name	chapter5_buildfirstmodel
☐ Owner	IAMR:Admin-OneClick
☐ Creation Time	Mon, 06.06.2022 20:13:28
☐ Model State	READY
☐ Training Job Status	MaxAutoMLJobRuntimeReached
☐ validation:f1_binary	0.906950
☐ Estimated Cost	7.563997
☐	
☐ TRAINING DATA:	
☐ Query	SELECT STATE, ACCOUNT_LENGTH, ARE...
☐	FROM CHAPTER5_BUILDFIRSTMODEL.C...
☐	WHERE RECORD_DATE < '2020-01-01'
☐ Target Column	CHURN
☐	
☐ PARAMETERS:	

Figure 5.7 – Model State READY

Apart from **Model State**, the SHOW MODEL command gives you other useful information about the model, for example, the query used, **Target Column**, **Model Type**, and **Function Name** to use when predicting. You can see that **Model Type** in our example is **xgboost**, which tells you that Amazon SageMaker has chosen the XGBoost algorithm to build the binary classification model:

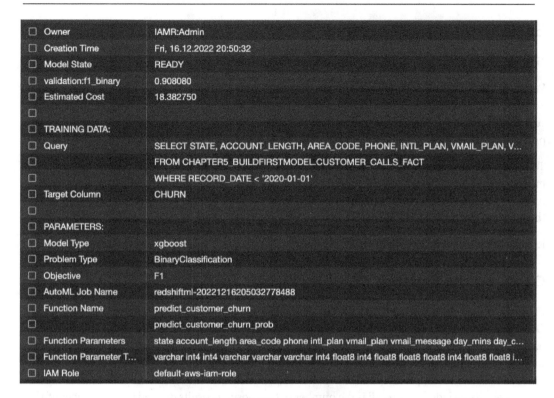

☐	Owner	IAMR:Admin
☐	Creation Time	Fri, 16.12.2022 20:50:32
☐	Model State	READY
☐	validation:f1_binary	0.908080
☐	Estimated Cost	18.382750
☐		
☐	TRAINING DATA:	
☐	Query	SELECT STATE, ACCOUNT_LENGTH, AREA_CODE, PHONE, INTL_PLAN, VMAIL_PLAN, V...
☐		FROM CHAPTER5_BUILDFIRSTMODEL.CUSTOMER_CALLS_FACT
☐		WHERE RECORD_DATE < '2020-01-01'
☐	Target Column	CHURN
☐		
☐	PARAMETERS:	
☐	Model Type	xgboost
☐	Problem Type	BinaryClassification
☐	Objective	F1
☐	AutoML Job Name	redshiftml-20221216205032778488
☐	Function Name	predict_customer_churn
☐		predict_customer_churn_prob
☐	Function Parameters	state account_length area_code phone intl_plan vmail_plan vmail_message day_mins day_c...
☐	Function Parameter T...	varchar int4 int4 varchar varchar varchar int4 float8 int4 float8 float8 float8 int4 float8 i...
☐	IAM Role	default-aws-iam-role

Figure 5.8 – Model State READY continuation

If you read further into the output, Redshift ML has done the bulk of the work for you, for example, it has selected and set the following parameters:

- **Problem Type** is set to **BinaryClassification**. This is true since our target variable has two distinct values in it, true and false. So, this is a binary classification problem.

- **Validation** and **Objective** is set to **F1**. F1 score is a recommended approach when evaluating binary scores since it considers both precision and recall. Other objectives that SageMaker Autopilot may select for a binary classification model are **accuracy** and **area under curve** (AUC).

We have created the model successfully as **Model State** shows as **READY**. The next step is to make use of prediction functions. We use them in SELECT queries. The next sections show how to do so.

Evaluating model performance

Now we have created the model, let's dive into the details of its performance.

When building machine learning models, it is very important to understand the model performance. You do this to make sure your model is useful and is not biased to one class over another and to make sure that the model is not under-trained or over-trained, which will mean the model is either not predicting classes correctly or is predicting only some instances and not others.

To address this problem, Redshift ML provides various objectives to measure the performance of the model. It is prudent that we test the model performance with the test dataset that we set aside in the previous section. This section explains how to review the Redshift ML objectives and also validate the model performance with our test data.

Redshift ML uses several objective methods to measure the predictive quality of machine learning models.

Checking the Redshift ML objectives

Figure 5.9 shows the SHOW MODEL output. It displays two values that are of interest to us. One is **Objective** and the other is **validation:f1_binary**. The first value to look at is **Objective**. It is set to **F1** for us. F1 or F-score is the most commonly used performance evaluation metric used for classification models. It is a measure for validating dataset accuracy. It is calculated from the precision and recall of the validations where precision is the number of true positive results divided by the number of all positive results included, and recall is the number of true positive results divided by the number of all records that should have been identified as positive. You can learn more about F-score here: https://en.wikipedia.org/wiki/F-score.

Run the following command in query editor v2:

```
SHOW MODEL chapter5_buildfirstmodel.customer_churn_model;
```

The output in *Figure 5.9* shows the value of F1 is found in **validation:f1_binary**, which is **0.90**. The highest possible value for an F1 score is 1 and the lowest is 0. The highest score of 1 would signify perfect precision and recall by a model. In our case, it is 90%, which is really good.

Key	Value
Model Name	customer_churn_model
Schema Name	chapter5_buildfirstmodel
Owner	IAMR:Admin
Creation Time	Fri, 16.12.2022 20:50:32
Model State	READY
validation:f1_binary	0.908080
Estimated Cost	18.382750
TRAINING DATA:	
Query	SELECT STATE, ACCOUNT_LEN...
	FROM CHAPTER5_BUILDFIRST...
	WHERE RECORD_DATE < '2020...
Target Column	CHURN
PARAMETERS:	
Model Type	xgboost
Problem Type	BinaryClassification
Objective	F1
AutoML Job Name	redshiftml-20221216205032778...
Function Name	predict_customer_churn

Figure 5.9 – Model objective values

We have seen that the model created by Autopilot has a good F-score and is ready to use to predict whether customers are going to churn or not. In the next section, we will use the prediction function to generate the prediction values along with probability scores.

Running predictions

Now let's invoke our `predict_customer_churn` and `predict_customer_churn_prob` prediction functions through the `SELECT` command. Redshift ML creates two functions for us to use:

- One is created with the same name as the one we gave when creating the model, in this case, `predict_customer_churn`, which returns the class label or predicted value, for example, 0 or 1.

- The other function, `predict_customer_churn_prob`, in addition to returning the class label or predicted value, also returns the probability that the predicted value is correct.

To test these functions, run the following query. In the following query, you'll notice that we are using two prediction functions inside a `SELECT` command and passing all the input columns that were passed when creating the ML model. These two functions will return a label and probability score as output. We are also testing the prediction function by filtering rows where `record_date` is greater than `'2022-07-31'`. Since this is an unseen dataset, it should act as a challenging dataset for our ML model.

It is also important to note that all the predictions are happening locally on a Redshift cluster. When the `SELECT` query is run, there are no calls made to Amazon SageMaker. This makes all predictions free of cost:

```
SELECT area_code ||phone  accountid, replace(churn,'.','') as Actual_
churn_class,
    chapter5_buildfirstmodel.predict_customer_churn(
      state,account_length,area_code, phone,intl_plan,
      vMail_plan, vMail_message, day_mins, day_calls,
      day_charge, total_charge, eve_mins, eve_calls,
      eve_charge, night_mins, night_calls,
      night_charge, intl_mins, intl_calls, intl_charge,
      cust_serv_calls) AS predicted_class,
    chapter5_buildfirstmodel.predict_customer_churn_prob(
      state, account_length, area_code, phone, intl_plan,
      vMail_plan, vMail_message, day_mins, day_calls,
      day_charge, total_charge, eve_mins, eve_calls,
      eve_charge, night_mins, night_calls,night_charge,
      intl_mins, intl_calls, intl_charge, cust_serv_calls)
      AS probability_score
  FROM chapter5_buildfirstmodel.customer_calls_fact
WHERE record_date > '2020-07-31'
  ;
```

You can see the output in the following screenshot:

accountid	actual_churn_class	predicted_class	probability_score
415382-4657	False	False	{"probabilities":[0.99597979,0.00402024],"labels":["False","True"]}
415358-1921	False	False	{"probabilities":[0.99215788,0.00784212],"labels":["False","True"]}
415330-6626	False	False	{"probabilities":[0.98575455,0.01424544],"labels":["False","True"]}
415329-9001	False	False	{"probabilities":[0.99430490,0.00569507],"labels":["False","True"]}
408335-4719	False	False	{"probabilities":[0.99906099,0.00093898],"labels":["False","True"]}
415330-8173	False	False	{"probabilities":[0.91623491,0.08376508],"labels":["False","True"]}
415351-7269	True	True	{"probabilities":[0.99852717,0.00147283],"labels":["True","False"]}
408350-8884	False	False	{"probabilities":[0.99868357,0.00131640],"labels":["False","True"]}
408393-7984	True	True	{"probabilities":[0.99582291,0.00417709],"labels":["True","False"]}
510343-4696	False	False	{"probabilities":[0.99293268,0.00706731],"labels":["False","True"]}
408418-6412	False	False	{"probabilities":[0.99657983,0.00342020],"labels":["False","True"]}
408383-1121	False	False	{"probabilities":[0.99966955,0.00033045],"labels":["False","True"]}
408360-1596	True	True	{"probabilities":[0.99825424,0.00174576],"labels":["True","False"]}
408395-2854	False	False	{"probabilities":[0.99945074,0.00054926],"labels":["False","True"]}
408341-9764	False	False	{"probabilities":[0.99818569,0.00181429],"labels":["False","True"]}
415353-3305	False	False	{"probabilities":[0.99953187,0.00046814],"labels":["False","True"]}
415402-1381	False	False	{"probabilities":[0.96224242,0.03775758],"labels":["False","True"]}

Figure 5.10 – Running predictions

In the preceding screenshot, observe that the predicted_class values and probability_score values for each customer are shown. From the predicted_class column, you can understand that our model is predicting whether the customer is going to churn or not, and from the probability_score column, you can understand that the model is, for example, for the first row, 99% confident that the customer with account ID **415382-4657** is not going to churn.

We have witnessed that prediction is working without any issues. In the next section, let's check how the model is performing compared to ground truth.

Comparing ground truth to predictions

Run the following query to compare actual versus predicted customer churn:

```
WITH infer_data AS (
  SELECT area_code ||phone   accounted,
    replace(churn,'.','') as churn,
    chapter5_buildfirstmodel.predict_customer_churn(
            state,
            account_length,
            area_code,
            phone,
```

```
                intl_plan,
                vMail_plan,
                vMail_message,
                day_mins,
                day_calls,
                day_charge,
                total_charge,
                eve_mins,
                eve_calls,
                eve_charge,
                night_mins,
                night_calls,
                night_charge,
                intl_mins,
                intl_calls,
                intl_charge,
                cust_serv_calls) AS predicted
   FROM chapter5_buildfirstmodel.customer_calls_fact
WHERE record_date > '2020-07-31'

)
SELECT *  FROM infer_data where churn!=predicted;
```

The following screenshot shows the customers where the ML model made a mistake:

> **Note**
> Results will vary as each trained model will have slight differences.

Result 1 (44)

accountid	churn	predicted
408370-7574	True	False
510395-3026	True	False
415334-1275	True	False
408339-3049	True	False
415402-5155	True	False
510334-9505	True	False
415405-3916	True	False
408339-7541	True	False
408393-7826	True	False
415410-6791	True	False
415400-6257	True	False
510401-8735	True	False
408368-1288	True	False
415362-5111	True	False

Figure 5.11 – Incorrect predictions

We have seen the model predictions and compared them with ground truth. In the next section, we will learn about feature importance.

Feature importance

Feature importance is a measure of how much each feature contributes to the model's predictions. SageMaker Autopilot calculates the importance of features and Redshift ML provides `explain_model` functions to retrieve feature importance. This will help you to understand which features are strongly related to the target variable, which features are important to the model and which are not, and from this you can reduce the number of dimensions that you feed into your machine learning model.

The following is the SQL code that you can run to retrieve the feature importance of our model:

```
Select jsondata.featureimp.explanations.kernel_shap.label0.global_
shap_values as value
from ( select explain_model( 'chapter5_buildfirstmodel.customer_churn_
model')as featureimp) jsondata ;
```

The following is the JSON format output of feature importance. You can read and understand the importance of each feature.

{"account_length":0.6010025166511206,"area_code":0.13957797386391988,"cust_serv_calls":0.8811129394075697,"day_calls":1.7294931653761064,
"day_charge":0.1307716966435415,"day_mins":0.5517368262792286,"eve_calls":0.4905554095206185,"eve_charge":0.12399856192909246,
"eve_mins":0.30232087934672688,"intl_calls":2.5018813460365956,"intl_charge":0.10320839870308187,"intl_mins":0.2851059575278758,
"intl_plan":6.241876667680464,"night_calls":0.6374489643584772,"night_charge":0.16776078144116883,"night_mins":0.4900811661911824,
"phone":0.03049082746947854,"state":0.033064217156854199,"total_charge":1.2856551438095108,"vmail_message":0.21522579719733876,"vmail_plan":0.08844318137511628}

Figure 5.12 – Feature importance raw output

For better readability of the feature importance, you may execute the following SQL code:

```
select t1.feature_imp, t1.value from
(
Select
'account_length' as feature_imp,
jsondata.featureimp.explanations.kernel_shap.label0.global_shap_
values.account_length as value
from ( select explain_model( 'chapter5_buildfirstmodel.customer_churn_
model')as featureimp) jsondata
union
select 'area_code' as feature_imp ,
jsondata.featureimp.explanations.kernel_shap.label0.global_shap_
values.area_code as value
from ( select explain_model( 'chapter5_buildfirstmodel.customer_churn_
model')as featureimp) jsondata
union
select 'cust_serv_calls' as feature_imp ,
jsondata.featureimp.explanations.kernel_shap.label0.global_shap_
values.cust_serv_calls as value
from ( select explain_model( 'chapter5_buildfirstmodel.customer_churn_
model')as featureimp) jsondata
union
select 'day_calls' as feature_imp ,
jsondata.featureimp.explanations.kernel_shap.label0.global_shap_
values.day_calls as value
from ( select explain_model( 'chapter5_buildfirstmodel.customer_churn_
model')as featureimp) jsondata
union
select 'day_charge' as feature_imp ,
jsondata.featureimp.explanations.kernel_shap.label0.global_shap_
values.day_charge as value
from ( select explain_model( 'chapter5_buildfirstmodel.customer_churn_
model')as featureimp) jsondata
union
select 'day_mins' as feature_imp ,
jsondata.featureimp.explanations.kernel_shap.label0.global_shap_
values.day_mins as value
```

```
from ( select explain_model( 'chapter5_buildfirstmodel.customer_churn_
model')as featureimp) jsondata
union
select 'eve_calls' as feature_imp ,
jsondata.featureimp.explanations.kernel_shap.label0.global_shap_
values.eve_calls  as value
from ( select explain_model( 'chapter5_buildfirstmodel.customer_churn_
model')as featureimp) jsondata
union
select 'eve_charge' as feature_imp ,
jsondata.featureimp.explanations.kernel_shap.label0.global_shap_
values.eve_charge  as value
from ( select explain_model( 'chapter5_buildfirstmodel.customer_churn_
model')as featureimp) jsondata
union
select 'eve_mins' as feature_imp ,
jsondata.featureimp.explanations.kernel_shap.label0.global_shap_
values.eve_mins  as value
from ( select explain_model( 'chapter5_buildfirstmodel.customer_churn_
model')as featureimp) jsondata
union
select 'intl_calls' as feature_imp ,
jsondata.featureimp.explanations.kernel_shap.label0.global_shap_
values.intl_calls   as value
from ( select explain_model( 'chapter5_buildfirstmodel.customer_churn_
model')as featureimp) jsondata
union
select 'intl_charge' as feature_imp ,
jsondata.featureimp.explanations.kernel_shap.label0.global_shap_
values.intl_charge    as value
from ( select explain_model( 'chapter5_buildfirstmodel.customer_churn_
model')as featureimp) jsondata
union
select 'intl_mins' as feature_imp ,
jsondata.featureimp.explanations.kernel_shap.label0.global_shap_
values.intl_mins     as value
from ( select explain_model( 'chapter5_buildfirstmodel.customer_churn_
model')as featureimp) jsondata
union
select 'intl_plan' as feature_imp ,
jsondata.featureimp.explanations.kernel_shap.label0.global_shap_
values.intl_plan      as value
from ( select explain_model( 'chapter5_buildfirstmodel.customer_churn_
model')as featureimp) jsondata
union
select 'inight_calls' as feature_imp ,
jsondata.featureimp.explanations.kernel_shap.label0.global_shap_
```

```
values.night_calls     as value
from ( select explain_model( 'chapter5_buildfirstmodel.customer_churn_
model')as featureimp) jsondata
union
select 'night_charge' as feature_imp ,
jsondata.featureimp.explanations.kernel_shap.label0.global_shap_
values.night_charge     as value
from ( select explain_model( 'chapter5_buildfirstmodel.customer_churn_
model')as featureimp) jsondata
union
select 'night_mins' as feature_imp ,
jsondata.featureimp.explanations.kernel_shap.label0.global_shap_
values.night_mins     as value
from ( select explain_model( 'chapter5_buildfirstmodel.customer_churn_
model')as featureimp) jsondata
union
select 'phone' as feature_imp ,
jsondata.featureimp.explanations.kernel_shap.label0.global_shap_
values.phone     as value
from ( select explain_model( 'chapter5_buildfirstmodel.customer_churn_
model')as featureimp) jsondata
union
select 'state' as feature_imp ,
jsondata.featureimp.explanations.kernel_shap.label0.global_shap_
values.state     as value
from ( select explain_model( 'chapter5_buildfirstmodel.customer_churn_
model')as featureimp) jsondata
union
select 'total_charge' as feature_imp ,
jsondata.featureimp.explanations.kernel_shap.label0.global_shap_
values.total_charge     as value
from ( select explain_model( 'chapter5_buildfirstmodel.customer_churn_
model')as featureimp) jsondata
union
select 'vmail_message' as feature_imp ,
jsondata.featureimp.explanations.kernel_shap.label0.global_shap_
values.vmail_message     as value
from ( select explain_model( 'chapter5_buildfirstmodel.customer_churn_
model')as featureimp) jsondata
union
select 'vmail_plan' as feature_imp ,
jsondata.featureimp.explanations.kernel_shap.label0.global_shap_
values.vmail_plan as value
from ( select explain_model( 'chapter5_buildfirstmodel.customer_churn_
model')as featureimp) jsondata
) t1
order by value desc
```

feature_imp	value
intl_plan	6.241876667680464
intl_calls	2.5018813460365956
day_calls	1.7294931653761064
total_charge	1.2856551438095108
cust_serv_calls	0.8811129394075697
inight_calls	0.6374489643584772
account_length	0.6010025166511206
day_mins	0.5517368262792286
eve_calls	0.49055554095206185
night_mins	0.4900811661911824
eve_mins	0.30232087934672688
intl_mins	0.2851059575278758
vmail_message	0.21522579719733876
night_charge	0.16776078144116883
area_code	0.13957797386391988
day_charge	0.1307716966435415
eve_charge	0.12399856192909246
intl_charge	0.10320839870308187
vmail_plan	0.08844318137511628
state	0.033064217156854199
phone	0.03049082746947854

Figure 5.13 – Feature Importance

You can use feature importance to understand the relationship between each feature and target variable and the features that are not important.

We have seen what features contribute highly to the model, now let's look at how model performance metrics are calculated on our test dataset.

Model performance

Let's use Redshift SQL to compute a **confusion matrix** to evaluate the performance of the classification model. Using a confusion matrix, you can identify true positives, true negatives, false positives, and false negatives, based on which various statistical measures such as accuracy, precision, recall, sensitivity, specificity, and finally, F1 score are calculated. You can read more about the concept of the confusion matrix here: https://en.wikipedia.org/wiki/Confusion_matrix.

The following query uses a WITH clause, which implements a common table expression in Redshift. This query has the following three parts:

- The first part is about the SELECT statement within the WITH clause, where we predict customer churn and save it in memory. This dataset is named infer_data.

- The second part, which is below the first SELECT statement, reads infer_data and builds the confusion matrix, and these details are stored in memory in a dataset called confusionmatrix.

- In the third part of the statement, note that the SELECT statement builds the model performance metrics such as F1 score, accuracy, recall, and so on.

Run the following query to build a confusion matrix for the test dataset:

```
WITH infer_data AS (
  SELECT area_code ||phone  accountid, replace(churn,'.','') as churn,
    chapter5_buildfirstmodel.predict_customer_churn(
        state,  account_length, area_code, phone,
        intl_plan, vMail_plan, vMail_message, day_mins,
        day_calls,  day_charge, total_charge, eve_mins,
        eve_calls, eve_charge, night_mins, night_calls,
        night_charge, intl_mins, intl_calls,
        intl_charge,  cust_serv_calls) AS predicted
  FROM chapter5_buildfirstmodel.customer_calls_fact
WHERE record_date > '2020-07-31'),
confusionmatrix as
(
SELECT case when churn  ='True' and predicted = 'True' then 1 else 0
end TruePositives,
case when churn ='False' and predicted = 'False' then 1 else 0 end
TrueNegatives,
case when churn ='False' and predicted = 'True' then 1 else 0 end
FalsePositives,
case when churn ='True' and predicted = 'False' then 1 else 0 end
FalseNegatives
  FROM infer_data
  )
select
```

```
sum(TruePositives+TrueNegatives)*1.00/(count(*)*1.00) as Accuracy,--
accuracy of the model,
sum(FalsePositives+FalseNegatives)*1.00/count(*)*1.00 as Error_Rate,
--how often model is wrong,
sum(TruePositives)*1.00/sum (TruePositives+FalseNegatives) *1.00 as
True_Positive_Rate, --or recall how often corrects are rights,
sum(FalsePositives)*1.00/sum (FalsePositives+TrueNegatives )*1.00
False_Positive_Rate, --or fall-out how often model said yes when it is
no,
sum(TrueNegatives)*1.00/sum (FalsePositives+TrueNegatives)*1.00 True_
Negative_Rate, --or specificity, how often model said no when it is
yes,
sum(TruePositives)*1.00 / (sum (TruePositives+FalsePositives)*1.00) as
Precision, -- when said yes how it is correct,
2*((True_Positive_Rate*Precision)/ (True_Positive_Rate+Precision) ) as
F_Score --weighted avg of TPR & FPR
From confusionmatrix
;
```

We get the following output:

Figure 5.14 – Confusion matrix for the test dataset

By looking at the **f_score** value, you can confirm that the model has performed well against our test dataset (`record_date > '2020-07-31'`). These records have not been seen by the model before, but 97% of the time, the model is able to correctly predict the class value. This proves that the model is useful and correctly predicts both classes – churn and no churn. This model can now be given to the business units so it can be used to proactively predict the customers who are about to churn and build marketing campaigns for them.

Summary

In this chapter, you have learned how to create your first machine learning model using a simple `CREATE MODEL` statement. While doing so, you explored `customer_calls_fact` table data using query editor v2, learned about the basic syntax of the `CREATE MODEL` statement, created a simple ML model, learned how to read the model's output, and finally, used Redshift SQL to compute some of the model evaluation metrics yourself.

In the next chapter, you will use the basics that you have learned in this chapter to build various classification models using Redshift ML.

6

Building Classification Models

In this chapter, you will learn about classification algorithms used in **machine learning** (**ML**). You will learn about the various methods that Redshift offers when you create classification models. This chapter will provide detailed examples of both **binary** and **multi-class classification models** and show you how to solve business problems with these modeling techniques. By the end of this chapter, you will be in a position to identify whether a business problem is a classification or not, identify the right method that Redshift offers in training, and build a model.

In this chapter, we will go through the following main topics:

- An introduction to classification algorithms
- Creating a model syntax with user guidance
- Training a binary classification model using the XGBoost algorithm
- Training a multi-class classification model using the Linear Learner model type

Technical requirements

This chapter requires a web browser and access to the following:

- An AWS account
- An Amazon Redshift Serverless endpoint
- Amazon Redshift Query Editor v2
- Completing the *Getting started with Amazon Redshift Serverless* section in *Chapter 1*

You can find the code used in this chapter here: `https://github.com/PacktPublishing/Serverless-Machine-Learning-with-Amazon-Redshift/`.

An introduction to classification algorithms

Classification is the process of categorizing any kind of entity or class so that it is better understood and analyzed. The classifying process usually happens as part of a pre-setup business process (for example, tagging a product as defective or good after observing it), or through a return process (for example, tagging a product as defective after the customer returned it as defective). In either event, the important point is classifying an entity – in this case, a product into a class (i.e., defective or not).

Figure 6.1 shows data that has been classified into two classes using three input variables. The figure shows where a pair of **Input** and **Output** data points are categorized into two classes. When output labels consist of only two classes, it is called a **binary classification** problem:

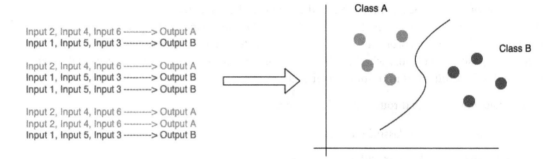

Figure 6.1 – Binary classification

If the output variable consists of more than two classes – for example, predicting whether a fruit is an apple, an orange, or a pear – then it is called **multi-class classification**. *Figure 6.2* shows data that has been classified into multiple classes based on a set of three input variables. The figure shows a multi-class classification chart, illustrating how input and output pairs are classified into three classes:

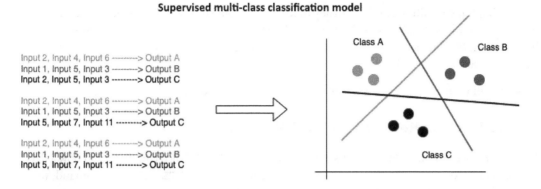

Figure 6.2 – Multi-class classification

The classification process can also happen on data that does not have classes defined yet. Let us continue to understand how this is possible.

It is not always the case that your entities are grouped or categorized in a certain way. For example, if you want to analyze your customers' purchase history or clickstream activity, or if you want to group similar customers based on demographics or shopping behavior, then classification algorithms come in handy to analyze the data and group similar data points into clusters. This type of classification modeling is called **unsupervised learning**.

Establishing classes helps the analysis process – for example, once products are tagged to a class label, you can easily retrieve a list of defective products that are returned and then further study the characteristics, such as store location, the demographics of the customer who returned the product, and the season when a product was returned most. How and when classes are defined and established enables businesses to conduct a deep-dive analysis, not only answering questions such as where and what but also training an ML model on historical data and classes, and predicting which class an entity will fall into.

Common use cases where classification models are useful include the following:

- Customer behavior prediction

- Document or image classification

- Spam filtering

In this chapter, we will show you how to create different classification models that Redshift offers you. Amazon Redshift provides **XGBoost**, **multilayer perceptron** (**MLP**), and **Linear Learner algorithms** to train and build a classification model.

In this chapter, you will begin the journey of learning about supervised classification models by building binary classification models, using XGBoost, and a multi-class classification model, using linear learner. MLP models will be covered in *Chapter 9*, whereas unsupervised classification modeling will be covered in *Chapter 8*.

Now, we will walk you through the detailed syntax of creating models with Redshift ML.

Diving into the Redshift CREATE MODEL syntax

In *Chapter 4*, we saw different variations of the Redshift CREATE MODEL command and how a data analyst, citizen data scientist, or data scientist can operate the CREATE MODEL command, with varying degrees of complexity. In this section, we will introduce you to a citizen data scientist persona, who is not fully aware of statistics but has good knowledge about identifying what algorithm to use and what problem type can be applied to a business problem. In the Redshift ML world, this type of model creation is known as **CREATE MODEL with user guidance**.

We are going to explore the model type and problem type parameters of the CREATE MODEL statement. As part of *CREATE MODEL with user guidance*, you also have the option of setting a preprocessor, but we will leave that topic for *Chapter 10*.

As an ML model creator, you will decide what algorithm to use and what problem type to address. Redshift ML still performs the feature engineering of independent variables behind the scenes. For example, out of 20 features, Redshift ML will automatically identify the categorical variables and numeric variables and create one-hot-encoded value or standardization of numerical variables where applicable, along with various other tasks required to complete the model training.

In summary, you let Redshift ML handle the bulk of data preparation tasks for ML. As a model creator, you come up with an algorithm to be used and a problem type to be solved. By preselecting an algorithm type and problem type, Redshift ML will reduce the training type, as it trains the model on other algorithms and problem types. Compared to the full AUTO CREATE MODEL statement that we created in *Chapter 5*, *CREATE MODEL with user guidance* takes less time.

As mentioned in the previous section, we will use the XGBoost algorithm for binary classification and the linear learner algorithm for multi-class classification.

You can learn more about XGBoost here: `https://docs.aws.amazon.com/sagemaker/latest/dg/XGBoost.html`.

And you can learn more about Linear Learner here: `https://docs.aws.amazon.com/sagemaker/latest/dg/linear-learner.html`.

Using a simple CREATE MODEL statement, Redshift ML will use SageMaker Autopilot to automatically determine the problem type, algorithm, and the best model type to use.

With Redshift ML, you can influence a model by providing user guidance. You can choose `model_type`, `problem_type`, and `objective` when you issue the `CREATE MODEL` statement. You can find more details on the syntax and options here: `https://docs.aws.amazon.com/redshift/latest/dg/r_create_model_use_cases.html`.

So far, we have discussed the basics of the Redshift ML `CREATE MODEL` syntax and how you can provide guidance, such as model type and objective, or choose to let Redshift ML automatically choose these for you.

Now, you will learn how to create a binary classification model and specify the XGBoost algorithm.

Training a binary classification model using the XGBoost algorithm

Binary classification models are used to solve the problem of predicting one class of two possible classes – for example, predicting whether it will rain or not. The goal is to learn about past data points and figure out which one of the target buckets a particular data point will fall into. The typical use cases of a binary classification model are as follows:

- Predicting whether a patient suffers from a disease
- Predicting whether a customer will churn or not
- Predicting behavior – for example, whether a customer will file an appeal or not

In the next few sections, we will go through the following steps to achieve our goal of creating a binary classification model to be used to run inference queries:

1. Defining the business problem
2. Uploading and analyzing data
3. Creating the model
4. Running prediction queries against the model

Establishing the business problem

To build our binary classification problem, we will take a look at a banking campaign issue. Banks spend a lot of money on marketing campaigns targeted toward their customers so that they will subscribe to their products. It is very important that banks build efficiency into their campaign, and this can be done by learning the last campaign dataset and predicting future campaign results. We will work on predicting whether a banking customer will subscribe to a banking product offer of a term deposit.

Uploading and analyzing the data

We are going to work on a bank marketing dataset in this section. The data is related to direct marketing campaigns of a Portuguese banking institution. Imagine you are a marketing analyst and your goal is to increase the amount of deposits by offering a term deposit to your customers. It is very important that marketing campaigns target customers appropriately. You will create a model using Redshift ML to predict whether a customer is likely to accept the term deposit offer. This dataset is sourced from `https://archive.ics.uci.edu/ml/datasets/bank+marketing`.

> **Dataset citation**
>
> [Moro et al., 2014] S. Moro, P. Cortez and P. Rita. *A Data-Driven Approach to Predict the Success of Bank Telemarketing. Decision Support Systems*, Elsevier, 62:22–31, June 2014

The classification goal is to predict whether the client will subscribe (yes/no) to a term deposit (the y variable).

The dataset has columns such as age, job, marital status, education level, and employment status.

Metadata about these columns can also be found at the UCI ML repository website here: `https://archive.ics.uci.edu/ml/datasets/bank+marketing`.

As you can see from the preceding link, there are 20 independent variables and 1 dependent variable (y). We can use any or all of these independent variables as input to our CREATE MODEL statement to be able to predict the outcome, y, which indicates whether the customer is likely to accept the offer.

After successfully connecting to Redshift as an admin or database developer, create the schema and load data into Amazon Redshift using the following steps:

1. Navigate to Redshift **query editor v2**, and connect to the **Serverless** endpoint and the **dev** database.

2. Rename the **Untitled** query editor by saving it as Chap6.

    The following screenshot shows the serverless connection, the database set to **dev**, and the query editor page saved as **Chap6**:

Figure 6.3 – Query Editor v2

3. Now, using the following line of code, create the schema. This schema is where all the tables and data needed for this chapter will be created and maintained:

```
Create schema chapter6_supervisedclassification;
```

You will see output like this, indicating that your schema is created:

Figure 6.4 – Schema created

The following code will create the `bank_details_training` table to store data to train the model, and the `bank_details_inference` table to store data to run the inference queries. Note that we have already split our input dataset into these two datasets for you. All of the SQL commands used in this chapter can be found here: `https://github.com/PacktPublishing/Serverless-Machine-Learning-with-Amazon-Redshift/blob/main/CodeFiles/chapter6/chapter6.sql`.

4. Run the following code from GitHub to create the training and inference tables in Query Editor v2:

```
CREATE TABLE chapter6_supervisedclassification.bank_details_
training(
    age numeric, "job" varchar marital varchar, education
varchar,
    "default" varchar, housing varchar, loan varchar,
    contact varchar, month varchar, day_of_week varchar,
    ...,
    y boolean ) ;

--create table to store data for running predictions

CREATE TABLE chapter6_supervisedclassification.bank_details_
inference(
    age numeric, "job" varchar marital varchar, education
varchar,
    "default" varchar, housing varchar, loan varchar,
    contact varchar, month varchar, day_of_week varchar,
    ...,

    y boolean ) ;
```

You will see output like this to verify that your tables have been created successfully:

Figure 6.5 – Tables created successfully

Now that you have created the tables, run the commands in *step 5* using Query Editor v2 to load the data, using the S3 buckets provided.

5. Load the sample data into the tables created in *step 4* by using the following command, which can be found on GitHub. Note that we use the COPY command to load this data from Amazon S3:

```
--load data into  bank_details_inference
TRUNCATE chapter6_supervisedclassification.bank_details_
inference;

 COPY chapter6_supervisedclassification.bank_details_inference
from 's3://packt-serverless-ml-redshift/chapter06/bank-
marketing-data/inference-data/inference.csv' REGION 'eu-west-1'
IAM_ROLE default CSV IGNOREHEADER 1 delimiter ';';
-- load data into bank_details_training
TRUNCATE chapter6_supervisedclassification.bank_details_
training;
 COPY chapter6_supervisedclassification.bank_details_training
from 's3://packt-serverless-ml-redshift/chapter06/bank-
marketing-data/training-data/training.csv' REGION 'eu-west-1'
IAM_ROLE default CSV IGNOREHEADER 1 delimiter ';';
```

6. Analyze the customer term deposit subscription table by creating a histogram chart. First, run the following command again using Query Editor v2:

```
SELECT y, COUNT(*) Customer_Count FROM chapter6_
supervisedclassification.bank_details_training
GROUP BY y
;
```

You can see in the result set that **36548** customers did not choose the bank's offer and **4640** did accept. You can also use the chart feature in Query Editor v2 to create a bar chart. Click on the **Chart** option found on the right-hand side in the **Result** pane:

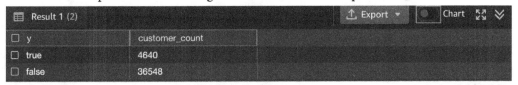

Figure 6.6 – The subscription results and the Chart option

You will get the following result after choosing **Bar** for **Type**, y for the **X** value, and **customer_count** for the **Y** value:

Figure 6.7 – A chart of customer acceptance

Now that we have our data loaded, we can create our model.

Using XGBoost to train a binary classification model

In this section, you will specify MODEL_TYPE and PROBLEM_TYPE to create a binary classification model using the XGBoost algorithm. We will now address the banking campaign problem. The goal of this model is to predict whether a customer will subscribe to a term deposit or not.

We will set MODEL_TYPE as XGBoost and PROBLEM_TYPE as BINARY_CLASSIFICATION. We will use the default IAM_ROLE. We also need to specify the S3 bucket where the model artifacts will be stored and, additionally, set MAX_RUNTIME to 3600 (in seconds).

The following is the code to create the model. You will find the complete code along with all the SQL commands needed for the chapter at https://github.com/PacktPublishing/Serverless-Machine-Learning-with-Amazon-Redshift/blob/main/chapter6.sql:

```
DROP MODEL chapter6_supervisedclassification.banking_termdeposit;

CREATE  MODEL chapter6_supervisedclassification.banking_termdeposit
FROM (
SELECT
    age ,
    "job" ,
    marital ,
    education ,
    "default" ,
    housing ,
    loan ,
    contact ,
    month ,
    day_of_week ,
    duration,
    campaign ,
    pdays ,
    previous ,
    poutcome ,
    emp_var_rate ,
    cons_price_idx ,
    cons_conf_idx ,
    euribor3m ,
    nr_employed ,
    y
FROM
    chapter6_supervisedclassification.bank_details_training )
    TARGET y
FUNCTION predict_term_deposit
IAM_ROLE default
MODEL_TYPE XGBoost
PROBLEM_TYPE BINARY_CLASSIFICATION
SETTINGS (
  S3_BUCKET '<<your-S3-Bucket',
  MAX_RUNTIME 9600
  )
;
```

By setting MODEL_TYPE to XGBoost and PROBLEM_TYPE to BINARY_CLASSIFICATION, we guide Redshift ML to only search for the best XGBoost model in this training run. If this is left as default, Redshift ML checks whether other classification models can be applied to the dataset.

Since the **SageMaker AutoPilot algorithm** does not have to test other model types or determine the problem type, the end result will be less training time. In this example, SageMaker Autopilot takes care of selecting the objective type, adjusting hyperparameters, and handling the data preprocessing steps.

To check the status of the model, run the following command:

```
SHOW MODEL chapter6_supervisedclassification.banking_termdeposit;
```

You will get the following result:

Figure 6.8 – Showing the model output

From the preceding screenshot, we can see that the model is still under training. Also, note that Redshift ML picks up the **Model Type** and **Problem Type** parameter values from our CREATE MODEL statement. Other parameters, such as the objective, hyperparameters, and preprocessing, are still auto-handled by Redshift ML.

The **predict_term_deposit** parameter under **Function Name** is used to generate predictions, which we will use in the next section.

Run the SHOW MODEL command again after some time to check whether model training is complete. From the following screenshot, you can see that **Model State** is **READY** and **F1** has been selected as the objective for model evaluation. The **F1** score is **0.646200**, or 64%. The closer this number is to 1, the better the model score:

Key	Value
Schema Name	chapter6_supervisedclassification
Owner	IAMR:Admin
Creation Time	Sun, 12.02.2023 19:02:25
Model State	READY
validation:f1_binary	0.646200
Estimated Cost	8.776864
TRAINING DATA:	
Query	SELECT AGE , "JOB" , MARITAL , EDUCATION , ...
	FROM CHAPTER6_SUPERVISEDCLASSIFICATIO...
Target Column	Y
PARAMETERS:	
Model Type	xgboost
Problem Type	BinaryClassification
Objective	F1
AutoML Job Name	redshiftml-20230212190225621701
Function Name	predict_term_deposit
	predict_term_deposit_prob
Function Parameters	age job marital education default housing loan co...
Function Parameter T...	numeric varchar varchar varchar varchar varchar ...

Figure 6.9 – Showing the model output

Let's run the following query against our training data to validate the F1 score:

```
WITH infer_data
 AS (
    SELECT  y as actual, chapter6_supervisedclassification.predict_
term_deposit(
    age ,    "job" ,    marital ,    education ,    "default" ,    housing
 ,   loan ,    contact ,    month ,    day_of_week ,    duration
 ,   campaign ,    pdays ,    previous ,    poutcome ,    emp_var_rate
 ,   cons_price_idx ,         cons_conf_idx ,         euribor3m ,    nr_
employed
) AS predicted,
    CASE WHEN actual = predicted THEN 1::INT
        ELSE 0::INT END AS correct
```

```
    FROM chapter6_supervisedclassification.bank_details_training
    ),
  aggr_data AS (
    SELECT SUM(correct) as num_correct, COUNT(*) as total FROM infer_
data
  )
  SELECT (num_correct::float/total::float) AS accuracy FROM aggr_data;
```

You can see in the following output that our accuracy is very good at almost 94%:

Figure 6.10 – The accuracy results

Now that the model training is complete, we will use the function created to run prediction queries.

Running predictions

Let us run some predictions on our inference dataset to see how many customers are predicted to subscribe to the term deposit. Run the following SQL statement in Query Editor v2:

```
WITH term_data AS ( SELECT chapter6_
supervisedclassification.predict_term_deposit( age,"job"
,marital,education,"default",housing,loan,contact,month,day_of_
week,duration,campaign,pdays,previous,poutcome,emp_var_rate,cons_
price_idx,cons_conf_idx,euribor3m,nr_employed) AS predicted
FROM chapter6_supervisedclassification.bank_details_inference )
SELECT
CASE WHEN predicted = 'Y'  THEN 'Yes-will-do-a-term-deposit'
    WHEN predicted = 'N'  THEN 'No-term-deposit'
    ELSE 'Neither' END as deposit_prediction,
COUNT(1) AS count
from term_data GROUP BY 1;
```

You should get the following output:

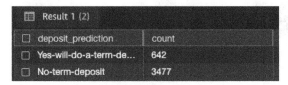

Figure 6.11 – Prediction results

We can see that **642** customers are predicted to accept the offer to subscribe to the term deposit, and **3477** are predicted to not accept the offer.

Prediction probabilities

Amazon Redshift ML now provides the capability to get the probability of a prediction for binary and multi-class classification problems. Note that in the output of the SHOW MODEL command in *Figure 6.9*, an additional function name has been created called predict_term_deposit_prob. Run the following query to check the probability that married customers with management jobs and between 35 and 45 years of age will accept the term deposit offer:

```
SELECT
age,"job" ,marital ,
chapter6_supervisedclassification.predict_term_deposit_prob( age,"job"
,marital,education,"default",housing,loan,contact,month,day_of_
week,duration,campaign,pdays,previous,poutcome,emp_var_rate,cons_
price_idx,cons_conf_idx,euribor3m,nr_employed) AS predicted
FROM chapter6_supervisedclassification.bank_details_inference
where marital = 'married'
  and "job" = 'management'
  and age between 35 and 40;
```

You will see the following results:

age	job	marital	predicted
36	management	married	{"probabilities":[0.99985629,0.00014372],"labels":["f","t"]}
40	management	married	{"probabilities":[0.99926740,0.00073259],"labels":["f","t"]}
35	management	married	{"probabilities":[0.83991116,0.16008882],"labels":["f","t"]}
39	management	married	{"probabilities":[0.94584262,0.05415738],"labels":["t","f"]}
37	management	married	{"probabilities":[0.52203691,0.47796309],"labels":["f","t"]}
37	management	married	{"probabilities":[0.55414915,0.44585085],"labels":["t","f"]}
36	management	married	{"probabilities":[0.75243974,0.24756023],"labels":["f","t"]}
35	management	married	{"probabilities":[0.98185575,0.01814427],"labels":["f","t"]}
38	management	married	{"probabilities":[0.55978036,0.44021964],"labels":["t","f"]}
39	management	married	{"probabilities":[0.99987215,0.00012783],"labels":["f","t"]}
40	management	married	{"probabilities":[0.89849180,0.10150822],"labels":["f","t"]}
36	management	married	{"probabilities":[0.88654792,0.11345207],"labels":["f","t"]}
35	management	married	{"probabilities":[0.78575808,0.21424192],"labels":["t","f"]}
38	management	married	{"probabilities":[0.99434894,0.00565108],"labels":["f","t"]}
36	management	married	{"probabilities":[0.99986333,0.00013665],"labels":["f","t"]}
38	management	married	{"probabilities":[0.99996585,0.00003415],"labels":["f","t"]}
39	management	married	{"probabilities":[0.91456062,0.08543941],"labels":["f","t"]}
35	management	married	{"probabilities":[0.59947753,0.40052247],"labels":["t","f"]}
36	management	married	{"probabilities":[0.99986386,0.00013614],"labels":["f","t"]}

Figure 6.12 – Probability results

You can see in the first row a **0.99985629** probability of a *false* prediction and only a **0.00014372** probability of a *true* prediction.

You can also modify the preceding query to see the probability of the customers that are predicted to accept the term deposit offer. Run the following SQL command in Query Editor v2:

```
SELECT age, "job", marital, predicted.labels[0], predicted.
probabilities[0]
from
  (select
age,"job" ,marital ,
chapter6_supervisedclassification.predict_term_deposit_prob( age,"job"
,marital,education,"default",housing,loan,contact,month,day_of_
week,duration,campaign,pdays,previous,poutcome,emp_var_rate,cons_
price_idx,cons_conf_idx,euribor3m,nr_employed) AS predicted
FROM chapter6_supervisedclassification.bank_details_inference
where marital = 'married'
   and "job" = 'management'
   and age between 35 and 40) t1
where predicted.labels[0] = 't';
```

You will see similar results as follows:

age	job	marital	labels	probabilities
39	management	married	"t"	0.94584262
37	management	married	"t"	0.55414915
35	management	married	"t"	0.59947753
38	management	married	"t"	0.66548312
37	management	married	"t"	0.99268746
38	management	married	"t"	0.66757184
37	management	married	"t"	0.82543176
35	management	married	"t"	0.78575808
38	management	married	"t"	0.83733559
36	management	married	"t"	0.88043034
38	management	married	"t"	0.55978036

Figure 6.13 – The probability results for customers accepting the term offer

In *Chapter 5*, you learned how to determine feature importance by running an explainability report. Run the following query to see which inputs contributed most to the model prediction:

```
select json_table.report.explanations.kernel_shap.label0.global_shap_
values from
  (select explain_model('chapter6_supervisedclassification.banking_
termdeposit') as report) as json_table;
```

Take the result and copy it to the editor so that it is easier to read, as shown in *Figure 6.14*:

{"age":0.5516878066167721,"campaign":0.2680577825694046,"cons_conf_idx":0.8824840903275205,"cons_price_idx":0.08693361607717634,"contact":0.2719735651883445,
"day_of_week":0.31254610183679767,"default":0.15016242273923889,"duration":8.814845209525759,"education":0.17914815472842664,"emp_var_rate":1.078969249528501,
"euribor3m":0.3777185610621967,"housing":0.09561040410322328,"job":0.2172898928675037,"loan":0.14497593955153463,"marital":0.1559494952507981 3,"month":0.8878134579089358,
"nr_employed":1.8428689683274489,"pdays":0.9377561028192128,"poutcome":0.05945855631250198,"previous":0.07209730914476756} |

Figure 6.14 – The explainability report

This shows that pdays has the most importance and that poutcome has the least.

Now that you have built a binary classification model, let us move on and try building a multi-class classification model.

Training a multi-class classification model using the Linear Learner model type

In this section, you will learn how to build a multi-class classification model in Amazon Redshift ML using the linear learner algorithm.

To do this, we will use a customer segmentation dataset from Kaggle: https://www.kaggle.com/datasets/vetrirah/customer.

You will use this dataset to train a model to classify customers into one of four segments (A, B, C, or D). By segmenting customers, you can better understand the customer and do targeted marketing to customers, with product offerings that are relevant to them.

Our data has already been split into training and testing sets and is stored in the following S3 locations:

- s3://packt-serverless-ml-redshift/chapter06/segmentation/train.csv

- s3://packt-serverless-ml-redshift/chapter06/segmentation/test.csv

After successfully connecting to Redshift as an admin or database developer, load data into Amazon Redshift as follows:

1. Navigate to Redshift **query editor v2**, and connect to the **Serverless** endpoint and the **dev** database.

2. Use the same schema and query editor page you created for the binary classification exercise (see the *Uploading and analyzing the data* section).

Create the train and test tables and load the data using the following SQL commands in Query Editor v2. These SQL commands can be found at https://github.com/PacktPublishing/Serverless-Machine-Learning-with-Amazon-Redshift/blob/main/CodeFiles/chapter6/chapter6.sql:

```
CREATE TABLE chapter6_supervisedclassification.cust_
segmentation_train (
    id numeric,
    gender varchar,
    ever_married  varchar,
    age numeric,
    graduated varchar,
    profession varchar,
    work_experience numeric,
    spending_score  varchar,
    family_size numeric,
    var_1 varchar,
    segmentation varchar
)
DISTSTYLE AUTO;


COPY chapter6_supervisedclassification.cust_segmentation_train
FROM 's3://packt-serverless-ml-redshift/chapter06/Train.
csv' IAM_ROLE DEFAULT FORMAT AS CSV DELIMITER ',' QUOTE '"'
IGNOREHEADER 1 REGION AS 'eu-west-1';


CREATE TABLE chapter6_supervisedclassification.cust_
segmentation_test (
    id numeric,
    gender varchar,
    ever_married  varchar,
    age numeric,
    graduated varchar,
    profession varchar,
    work_experience numeric,
    spending_score  varchar,
    family_size numeric,
    var_1 varchar
)
DISTSTYLE AUTO;


COPY chapter6_supervisedclassification.cust_segmentation_test
FROM 's3://packt-serverless-ml-redshift/chapter06/Test.csv' IAM_
ROLE DEFAULT FORMAT AS CSV DELIMITER ',' QUOTE '"' IGNOREHEADER
1 REGION AS 'eu-west-1';
```

Now that the data has loaded, let's do some analysis of our training data.

3.   Analyze the training data by executing the following SQL command:

```
select segmentation, count(*)  from chapter6_
supervisedclassification.cust_segmentation_train
group by 1;
```

You should get the following output:

☐ segmentation	count
☐ C	1970
☐ D	2268
☐ B	1858
☐ A	1972

Figure 6.15 – Segmentation

Our training dataset has a total of 8,068 customer records. From this sample, we can see that segments **C, B**, and **A** are very similar and that more customers are in segment **D**.

We will use the input from the training dataset to predict the customer segment, using the linear learner algorithm.

Using Linear Learner to predict the customer segment

Linear learner is a supervised learning algorithm and one of the model types you can use to solve classification or regression problems.

For multi-class classification problems, we have more than two labels (or targets) that we will try to predict, compared to exactly two labels for binary classification problems. We will show you how to use linear learner to solve regression problems in *Chapter 7*.

With linear learner, you can achieve a significant increase in speed compared to traditional hyperparameter optimization techniques, making it very convenient.

We will provide a training set with data that contains our input or observations about the data, and the label that represents the value we want to predict. We can optionally provide certain combinations of preprocessors to certain sets of columns.

In this section, you will apply user guidance techniques by providing MODEL_TYPE, PROBLEM_TYPE, and OBJECTIVE to create a multi-class classification model using the linear learner algorithm. The goal of this model is to predict the segment for each customer.

We will set MODEL_TYPE as LINEAR_LEARNER and PROBLEM_TYPE as MULTICLASS_CLASSIFICATION. We will leave other options as default.

Let us execute the following code in Query Editor v2 to train the model:

```
CREATE  MODEL chapter6_supervisedclassification.cust_segmentation_
model_11
FROM (
SELECT
    id, gender, ever_married, age, graduated,profession,
    work_experience, spending_score,family_size,
    var_1,segmentation
FROM chapter6_supervisedclassification.cust_segmentation_train
)
TARGET segmentation
FUNCTION predict_cust_segment_11    IAM_ROLE default
MODEL_TYPE LINEAR_LEARNER
PROBLEM_TYPE MULTICLASS_CLASSIFICATION
OBJECTIVE 'accuracy'
SETTINGS (
  S3_BUCKET '<<your-s3-bucket>>',
  S3_GARBAGE_COLLECT OFF,
  MAX_RUNTIME 9600
  );
```

To check the status of the model, run the following command in Query Editor v2:

```
SHOW MODEL chapter6_supervisedclassification.cust_segmentation_model_
11;
```

You should get the following output:

Key	Value
Model Name	cust_segmentation_model_ll
Schema Name	chapter6_supervisedclassification
Owner	IAMR:Admin
Creation Time	Tue, 14.02.2023 14:04:46
Model State	READY
validation:multiclass_...	0.535028
Estimated Cost	3.154642
TRAINING DATA:	
Query	SELECT ID, GENDER, EVER_MARRIED, AGE, GRADUATED,PROFESSION, WORK_EXPERIENCE, SPENDING_SCOR...
	FROM CHAPTER6_SUPERVISEDCLASSIFICATION.CUST_SEGMENTATION_TRAIN
Target Column	SEGMENTATION
PARAMETERS:	
Model Type	linear_learner
Problem Type	MulticlassClassification
Objective	Accuracy
AutoML Job Name	redshiftml-20230214140446461427
Function Name	predict_cust_segment_ll
	predict_cust_segment_ll_prob

Figure 6.16 – Showing the model output

You can see that the model is now in the **READY** state and that Redshift ML picks up **Model Type** and **Problem Type** parameter values from our CREATE MODEL statement.

Now that the model is trained, it is time to evaluate its quality.

Evaluating the model quality

When you issue the CREATE MODEL command, Amazon SageMaker will automatically divide your data into testing and training in the background so that it can determine the accuracy of the model. If you look at the validation:multiclass_accuracy key from the SHOW MODEL output, you will see a value of **0.535028**, which means our model can correctly pick the segment 53% of the time. Ideally, we prefer a value closer to 1.

We can also run a validation query to check our accuracy rates. In the following query, note that we select the actual segmentation, and then we use the function that was generated by our CREATE MODEL command to get the predicted segmentation to do the comparison:

```
select
cast(sum(t1.match)as decimal(7,2)) as predicted_matches
,cast(sum(t1.nonmatch) as decimal(7,2)) as predicted_non_matches
,cast(sum(t1.match + t1.nonmatch) as decimal(7,2))  as total_
predictions
```

```
,predicted_matches / total_predictions as pct_accuracy
from
(SELECT
    id,
    gender,
    ever_married,
    age,
    graduated,
    profession,
    work_experience,
    spending_score,
    family_size,
    var_1,
    segmentation as actual_segmentation,
    chapter6_supervisedclassification.predict_cust_segment_ll
(id,gender,ever_married,age,graduated,profession,work_experience,
spending_score,family_size,var_1) as predicted_segmentation,
    case when actual_segmentation = predicted_segmentation then 1
        else 0 end as match,
  case when actual_segmentation <> predicted_segmentation then 1
        else 0 end as nonmatch

    FROM chapter6_supervisedclassification.cust_segmentation_train
) t1;
```

We get the following output:

predicted_matches	predicted_non_matches	total_predictions	pct_accuracy
4291	3777	8068	0.53185423

Figure 6.17 – The model accuracy

This output shows that we are very close to the score of **.535028** when we compare the number of times the model correctly predicted the segment against the total number of input records.

Now that we have checked the model accuracy, we are ready to run prediction queries against the test dataset.

Running prediction queries

Now that we have our model and have done validation, we can run our prediction query against our test set so that we can segment our prospective customers, based on customer IDs. You can see that we now use our function against the test table to get the predicted segment:

```
SELECT
id,
chapter6_supervisedclassification.predict_cust_segment_11
(id,gender,ever_married,age,graduated,profession,work_
experience,spending_score,family_size,var_1) as  segmentation
FROM chapter6_supervisedclassification.cust_segmentation_test;
```

The first 10 customers are shown here:

□ id	segmentation
□ 458989	A
□ 458994	C
□ 458996	A
□ 459000	B
□ 459001	D
□ 459003	B
□ 459005	C
□ 459008	C
□ 459013	C
□ 459014	D

Figure 6.18 – The predicted segment

Let's see how the new prospective customers are spread across the various segments:

```
SELECT
    chapter6_supervisedclassification.predict_cust_segment_11
    (id,gender,ever_married,age,graduated,profession,work_
experience,spending_score,family_size,var_1) as  segmentation,
    count(*)
    FROM chapter6_supervisedclassification.cust_segmentation_test
    group by 1;
```

We can see here how many prospective customers are in each segment:

Result 1 (4)

segmentation	count
A	782
C	583
D	749
B	513

Figure 6.19 – The customer count by segment

Now that you have this information, your marketing team is ready to target their efforts on these prospective customers.

Let's now take a look at some other options you can use to solve this multi-class classification problem.

Exploring other CREATE MODEL options

We can also create this model in a couple of different ways, which we will explore in the following sections. It is important to understand the different options available so that you can experiment and choose the approach that gives you the best model.

In the first example, we will not provide any user guidance, such as specifying `MODEL_TYPE`, `PROBLEM_TYPE`, or `OBJECTIVE`. Use this approach if you are new to ML and want to let SageMaker Autopilot determine this for you.

Then, in the next example, you can see how you can provide `PROBLEM_TYPE` and `OBJECTIVE`. As a more experienced user of ML, you should now recognize which `PROBLEM_TYPE` and `OBJECTIVE` instances are best for your use case. When you provide these inputs, it will speed up the model training process, since SageMaker Autopilot will only train using the provided user guidance.

Creating a model with no user guidance

In this approach, we let SageMaker Autopilot choose `MODEL_TYPE`, `PROBLEM_TYPE`, and `OBJECTIVE`:

```
CREATE MODEL chapter6_supervisedclassification.cust_segmentation_model
FROM (
SELECT
    id,
    gender,
    ever_married,
    age,
    graduated,
    profession,
```

```
    work_experience,
    spending_score,
    family_size,
    var_1,
    segmentation
FROM chapter6_supervisedclassification.cust_segmentation_train
)
TARGET segmentation
FUNCTION predict_cust_segment   IAM_ROLE default
SETTINGS (
  S3_BUCKET '<<your S3 Bucket>>',
  S3_GARBAGE_COLLECT OFF,
  MAX_RUNTIME 9600
);
```

Note that we have only provided the basic settings. We did not specify MODEL_TYPE, PROBLEM_
TYPE, or OBJECTIVE. Amazon Redshift ML and SageMaker will automatically figure out that this
is a multi-class classification problem and use the best model type. As an additional exercise, run
this CREATE MODEL command, and then run the SHOW MODEL command. It will show you the
MODEL_TYPE parameter that Amazon SageMaker used to train the model.

Creating a model with some user guidance

In this example, we will provide PROBLEM_TYPE and OBJECTIVE, but we will let Amazon SageMaker
determine MODEL_TYPE:

```
CREATE MODEL chapter6_supervisedclassification.cust_segmentation_
model_ug
FROM (
SELECT
    id,
    gender,
    ever_married,
    age,
    graduated,
    profession,
    work_experience,
    spending_score,
    family_size,
    var_1,
    segmentation
FROM chapter6_supervisedclassification.cust_segmentation_train
)
TARGET segmentation
```

```
FUNCTION predict_cust_segment_ug   IAM_ROLE default
PROBLEM_TYPE MULTICLASS_CLASSIFICATION
OBJECTIVE 'accuracy'
SETTINGS (
  S3_BUCKET '<<your S3 Bucket>>',
  S3_GARBAGE_COLLECT OFF,
  MAX_RUNTIME 9600
  );
```

In this example, we let Amazon Redshift ML and Amazon SageMaker determine MODEL_TYPE, and we pass in PROBLEM_TYPE and OBJECTIVE. When you have some free time, experiment with the different methods of creating the models, and note the differences you see in the time it takes to train the model, and also compare the accuracy and other outputs of the SHOW MODEL command.

You can also create multi-class classification models using XGBoost, which we will cover in *Chapter 10*.

Summary

In this chapter, we discussed classification models in detail and looked at their common use cases. We also explained the CREATE MODEL syntax for classification models, where you provide guidance to train a model by supplying the model type and objective.

You learned how to do binary classification and multi-class classification with Amazon Redshift ML and how to use the XGBoost and linear learner algorithms. We also showed you how to check the status of your models, validate them for accuracy, and write SQL queries to run predictions on your test dataset.

In the next chapter, we will show you how to build regression models using Amazon Redshift ML.

7

Building Regression Models

In the previous chapter, we learned about classification models. In this chapter, we will learn about building **linear regression** models where we predict numeric variables. Unlike classification models, linear regression models are used to predict a **continuous numeric value**. Similar to the previous chapter, here also you will learn about various methods that Redshift provides for creating linear regression models.

This chapter will provide several detailed examples of business problems that can be solved with these modeling techniques. In this chapter, we will walk through how you can try different algorithms to get the best regression model.

By the end of this chapter, you will be in a position to identify whether a business problem is a linear regression or not and then be able to identify the right method that Redshift provides to train and build the model.

In this chapter, we will go through the following main topics:

- Introducing regression algorithms
- Creating a simple regression model using the XGBoost algorithm
- Creating multi-input regression models

Technical requirements

This chapter requires a web browser and access to the following:

- AWS account
- Amazon Redshift Serverless endpoint
- Amazon Redshift Query Editor v2

You can find the code used in this chapter here:

```
https://github.com/PacktPublishing/Serverless-Machine-Learning-with-
Amazon-Redshift/blob/main/CodeFiles/chapter7/chapter7.sql
```

Introducing regression algorithms

Regression models are used where you are trying to predict a numeric outcome such as what price an item will sell for. The outcome variable is your target and your input variables are used to determine the relationship between the variables so that you can predict the unknown target on sets of data without the target variable.

You can have a single input variable, also known as **simple linear regression**. For example, years of experience and salary usually have a relationship.

Multiple linear regression is when you have multiple input variables. For example, predicting the selling price of homes in a particular zip code by using the relationship between the target (price) and various inputs such as square footage, number of bedrooms, pool, basement, lot size, and year built.

A good linear regression model has a small amount of vertical distance between the line and the data points. Refer to the following figure:

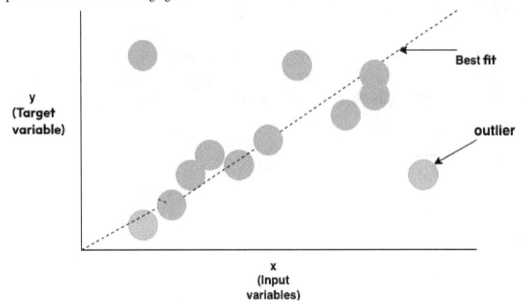

Figure 7.1 – Linear regression line

Common use cases where regression models are useful are as follows:

- Price and revenue prediction
- Predicting customer lifetime value
- Predicting the weather
- Measuring the effectiveness of marketing campaigns

This chapter will show you how to build regression models in Amazon Redshift ML using the XGBoost and Linear Learner algorithms, which you used in *Chapter 6*. As you will see, you can use the same algorithms on different machine learning problems.

We have looked at the regression problem; now let's look at the Redshift CREATE MODEL command to create a regression model.

Redshift's CREATE MODEL with user guidance

When using the CREATE MODEL command in Redshift, the system will automatically search for the best combination of preprocessing and model for your specific dataset. However, in some cases, you may want additional control over the model creation process or to incorporate domain-specific knowledge.

Redshift offers flexibility to guide the CREATE MODEL process so the time taken by the AutoML job is reduced.

We are going to explore the model type and problem type parameters of the CREATE MODEL statement in this chapter. As part of CREATE MODEL with user guidance, you also have the option of setting a preprocessor, but we will leave that topic for *Chapter 10*.

When you are guiding the AutoML job, as a machine learning model creator, you are going to decide what algorithm to use and what problem type to address. Redshift ML still performs the feature engineering of independent variables behind the scenes – for example, out of 20 features, Redshift ML will automatically identify the categorical variables and numeric variables and create a one-hot-encoded value or standardization of numerical variables where applicable, along with various other tasks required to complete the model training.

In summary, you are letting Redshift ML handle the bulk of data preparation tasks for machine learning. As a model creator, you have the option to specify the algorithm and problem type to be used in the CREATE MODEL statement, which has the benefit of reduced training time, since SageMaker does not need to spend time determining which algorithm or problem type to use.

Now that we have learned what CREATE MODEL with user guidance is, let's start creating a simple regression model.

Creating a simple linear regression model using XGBoost

To build our simple linear regression problem, we are going to take a look at a dataset that includes predicting weight based on height. This dataset has only one independent variable, which is height in inches, and is used to predict weight in pounds. Since there is only one independent variable, we call this the **simple linear regression** model.

In this section, we will upload the data, analyze it, prepare it for training the model, and then lastly, we will create the model and run prediction queries using the function created by the model.

Uploading and analyzing the data

We are going to work on a height and weight dataset in this section.

The dataset contains 25,000 synthetic records of human heights and weights of 18-year-old participants. This data was generated based on a 1993 Growth Survey, which was conducted on 25,000 children from their birth to 18 years of age. The participants were recruited from **Maternal and Child Health Centers** (**MCHCs**) and schools, and the data collected was used to develop Hong Kong's current growth charts for weight, height, weight-for-age, weight-for-height, and **body mass index** (**BMI**).

More details about this dataset can be found here: `http://wiki.stat.ucla.edu/socr/index.php/SOCR_Data_Dinov_020108_HeightsWeights`.

Dataset citation

Hung-Kwan So, Edmund AS Nelson, Albert M Li, Eric MC Wong, Joseph TF Lau, Georgia S Guldan, Kwok-Hang Mak, Youfa Wang, Tai-Fai Fok, and Rita YT Sung. (2008) *Secular changes in height, weight, and body mass index in Hong Kong Children*. BMC Public Health. 2008; 8: 320. doi: 10.1186/1471-2458-8-320. PMCID: PMC2572616

Leung SS, Lau JT, Tse LY, Oppenheimer SJ. *Weight-for-age and weight-for-height references for Hong Kong children from birth to 18 years*. J Paediatr Child Health. 1996;32:103–109. doi: 10.1111/j.1440-1754.1996.tb00904.x.

Leung SS, Lau JT, Xu YY, Tse LY, Huen KF, Wong GW, Law WY, Yeung VT, Yeung WK, et al. *Secular changes in standing height, sitting height and sexual maturation of Chinese–the Hong Kong Growth Study, 1993*. Ann Hum Biol. 1996;23:297–306. doi: 10.1080/03014469600004532.

In the following subsections, we will discuss the prediction goals we are trying to achieve using this dataset and then analyze the data.

Prediction goal

The goal is to predict the weight of children as a numeric value based on supplied height as a numeric value.

The dataset has the following columns:

Column	Description
Index	Sequential number
Height in Inches	Height of a child as a numerical value
Weight in Pounds	Weight of a child as a numerical value

Table 7.1 – Data definition

Analyzing the data

After successfully connecting to Redshift as an admin or database developer, create the schema and load data into Amazon Redshift as follows:

1. Navigate to **Redshift query editor v2** and connect to the **Serverless** endpoint and then the **dev** database.

2. Name the untitled query editor by saving it as Chapter7.

    The following screenshot shows a serverless connection, the database set to **dev**, and the **Redshift query editor** page saved as Chapter7:

Figure 7.2 – Connecting to the Serverless endpoint

3. Create the schema as follows:

    ```
    create schema chapter7_RegressionModel;
    ```

4. Create a table using the following code:

    ```
    --create table to load data
    DROP TABLE chapter7_RegressionModel.height_weight;
    ```

```
CREATE TABLE chapter7_RegressionModel.height_weight
(
    Id integer,
    HeightInches decimal(9,2),
    weightPounds decimal(9,2)
)
;
```

5. Load the sample data by using the following command:

```
TRUNCATE chapter7_RegressionModel.height_weight;
COPY chapter7_RegressionModel.height_weight
FROM 's3://packt-serverless-ml-redshift/chapter07/heightweight/
HeightWeight.csv'
IAM_ROLE default
CSV
IGNOREHEADER 1
REGION AS 'eu-west-1';
```

6. Analyze the height and weight dataset table by creating a histogram chart.

7. Use the Query Editor v2 **Chart** feature to create a graph. First, run the following command and then click on the **Chart** option found on the right-hand side in the **Results** pane:

```
SELECT * FROM
chapter7_RegressionModel.height_weight
ORDER BY 2,3;
```

To generate the following chart, you need to add two traces to your chart. By default, the chart is loaded with one trace, so you need to add one additional trace. You can add it by clicking on the + **Trace** button.

The following chart shows both variables. For *trace 1*, select **heightinches** on the y axis, leaving the x axis empty. For *trace 2*, select **weightpounds** on the y axis, leaving the x axis empty. The resulting chart should look like this:

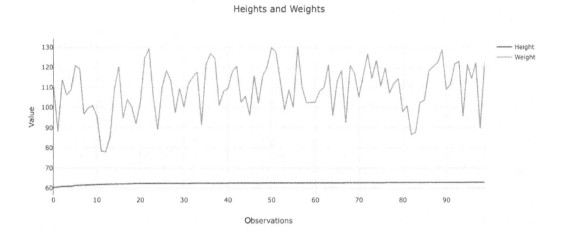

Figure 7.3 – Weights and heights

As you can see, there is a slight upward trend in weights along with heights.

Now that we have analyzed our dataset, we will split it into training and validation sets. The training set will be used to train our model and the validation set will be used to evaluate the model's accuracy.

Splitting data into training and validation sets

Since we have only one dataset, let's write a query that splits data into two logical sets: training and validation.

To train the model, let's use the syntax *where id%8 is not equal to 0*:

```
SELECT * FROM
chapter7_RegressionModel.height_weight Where id%8!=0;
```

To validate the model, let's use *where id%8 is equal to 0*:

```
SELECT * FROM
chapter7_RegressionModel.height_weight Where id%8=0;
```

We have analyzed and prepared our input data, now let's create a machine learning model.

Creating a simple linear regression model

In this section, you will use CREATE MODEL with user guidance to create a simple linear regression model using the XGBoost algorithm. We will address the weight prediction problem by training a machine learning model. The goal of this model is to predict a weight based on a given height.

We set MODEL_TYPE as xgboost and PROBLEM_TYPE as regression. We leave other options as default:

```
DROP MODEL chapter7_RegressionModel.predict_weight;
CREATE MODEL chapter7_RegressionModel.predict_weight
FROM (select heightinches, cast(round(weightpounds,0) as integer)
weightpounds from chapter7_RegressionModel.height_weight where id%8!=0
)
TARGET weightpounds
FUNCTION predict_weight
IAM_ROLE default
MODEL_TYPE xgboost
PROBLEM_TYPE regression
OBJECTIVE 'mse'
SETTINGS (s3_bucket '<<your-S3-bucket>>',
         s3_garbage_collect off,
         max_runtime 3600);
```

Let's take a look at the options we provided in the CREATE MODEL statement and discuss how they affect the actions taken by Amazon SageMaker

In the CREATE MODEL statement, we are guiding Redshift ML to use XGBoost as an algorithm by setting MODEL_TYPE. The Amazon SageMaker Autopilot job will not train the model using other supported algorithms – for example, **Linear Learner** or **multilayer perceptron** (**MLP**). When this option is left as default, Amazon SageMaker will train the model using all the algorithms supported by Autopilot.

Next, when we set PROBLEM_TYPE to regression, we are guiding Redshift ML to search for a model to solve a regression problem type.

We set OBJECTIVE to mse (**mean squared error**), which is commonly used to evaluate the performance of a regression model. It is a measure of the average of the squared differences between the predicted values and the actual values.

With these three guiding options, we are creating boundaries for Amazon SageMaker Autopilot. The end result would be less training time bundled with other benefits of the Autopilot algorithm – for example, adjusting hyperparameters and data preprocessing steps, which are all auto-handled by Amazon SageMaker Autopilot.

To check the status of the model, run the following command:

```
SHOW MODEL chapter7_RegressionModel.predict_weight;
```

The following screenshot shows the output of the SHOW MODEL command:

Key	Value
Model Name	predict_weight
Schema Name	chapter7_regressionmodel
Owner	IAMR:Admin
Creation Time	Wed, 15.02.2023 13:07:52
Model State	TRAINING
TRAINING DATA:	
Query	SELECT HEIGHTINCHES, CAST(ROUND(WEIGHTPOUNDS,0) AS I...
	FROM CHAPTER7_REGRESSIONMODEL.HEIGHT_WEIGHT
	WHERE ID%8!=0
Target Column	WEIGHTPOUNDS
PARAMETERS:	
Model Type	xgboost
Problem Type	Regression
Objective	MSE
AutoML Job Name	redshiftml-20230215130752296401
Function Name	predict_weight
Function Parameters	heightinches
Function Parameter T...	numeric

Figure 7.4 – SHOW MODEL output

The model is still under training, but you will notice that Redshift ML is picking up **Model Type**, **Problem Type**, and **Objective** parameter values from our CREATE MODEL statement.

The **Function Name** parameter, predict_weight, is used to generate predictions and is used in the SELECT statement, which we will cover in the next section.

Run the SHOW MODEL command again after some time to check whether the model training is complete or not. From the following screenshot, you can see that model training has finished and MSE has been selected as the objective for model evaluation. This is auto-selected by the Redshift ML and is the correct evaluation method for linear regression models:

Key	Value
Model Name	predict_weight
Schema Name	chapter7_regressionmodel
Owner	IAMR:Admin-OneClick
Creation Time	Wed, 27.07.2022 17:48:25
Model State	READY
validation:mse	102.496033
Estimated Cost	4.990489
TRAINING DATA:	
Query	SELECT HEIGHTINCHES, CAST(ROUND(WEIGHTPOUNDS,0) AS INTEGER) ...
	FROM CHAPTER7_REGRESSIONMODEL.HEIGHT_WEIGHT
	WHERE ID%8!=0
Target Column	WEIGHTPOUNDS
PARAMETERS:	
Model Type	xgboost
Problem Type	Regression
Objective	MSE
AutoML Job Name	redshiftml-20220727174825491549
Function Name	predict_weight
Function Parameters	heightinches
Function Parameter T...	numeric
IAM Role	default-aws-iam-role
S3 Bucket	serverlessmachinelearningwithredshift-709512860261
Max Runtime	3600

Figure 7.5 – SHOW MODEL output – model ready state

We have trained and created the model; in the next step, we will generate the predictions.

Running predictions

Since our model has been successfully trained, let's run some predictions against unseen datasets.

Run the following query to find records where the model is exactly predicting weight in pounds for a given height in inches where id%8=0. By using WHERE id%8=0, we are looking at ~20% of our dataset. These are records that were not included in model training. If you recall, in the CREATE MODEL statement, we specified WHERE id%8 !=0:

```
SELECT heightinches, CAST(chapter7_RegressionModel.predict_
weight(CAST(ROUND(heightinches,0) as integer)) as INTEGER) as
Predicted_Weightpounds,
   CAST(ROUND(weightpounds,0) as INTEGER) Original_Weightpounds ,
```

```
Predicted_Weightpounds - Original_Weightpounds  as Difference
FROM chapter7_RegressionModel.height_weight WHERE id%8=0
AND Predicted_Weightpounds - Original_Weightpounds = 0;
```

Here is the output for it:

heightinches	predicted_weightpounds	original_weightpounds	difference
65.81	121	121	0
65.31	116	116	0
67.29	124	124	0
68.57	130	130	0
68.8	130	130	0
72.56	143	143	0
66.88	124	124	0

Result 1 (100)

Figure 7.6 – Showing predicted weight results

Now, let's check the MSE and **root mean square error** (**RMSE**):

```
SELECT
  ROUND(AVG(POWER(( Original_Weightpounds - Predicted_Weightpounds
),2)),2) mse

  , ROUND(SQRT(AVG(POWER(( Original_Weightpounds - Predicted_
Weightpounds ),2))),2) rmse
FROM
  ( select heightinches, cast(chapter7_RegressionModel.predict_
weight(cast(round(heightinches,0) as integer)) as integer) as
Predicted_Weightpounds,
  cast(round(weightpounds,0) as integer) Original_Weightpounds ,
  Predicted_Weightpounds - Original_Weightpounds as Difference
  from chapter7_RegressionModel.height_weight where id%8=0
);
```

Here is the output:

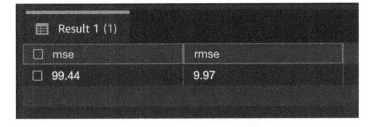

mse	rmse
99.44	9.97

Result 1 (1)

Figure 7.7 – MSE and RMSE values

Our MSE value is high; it represents data that may have outliers or for which we do not have enough variables. For example, adding age and gender may improve the prediction score.

Let's compare predicted scores and original scores in a line chart:

```
select heightinches, cast(chapter7_RegressionModel.predict_
weight(cast(round(heightinches,0) as integer)) as integer) as
Predicted_Weightpounds,
   cast(round(weightpounds,0) as integer) Original_Weightpounds ,
   Predicted_Weightpounds - Original_Weightpounds as Difference
   from chapter7_RegressionModel.height_weight where id%8=0;
```

Once a response is returned, click on the **Chart** option found on the right-hand side in the Query Editor, add a trace for the line, and select **Predicted_Weightpounds**. Add another trace for the line chart and select **Original_Weightpounds**, then add a third trace, but this time, select **Bar graph** and add a **Difference** column.

In the following chart, you will notice that the predicted scores are following the original scores. The difference is shown at the bottom of the graph, which gives information about the variance or error:

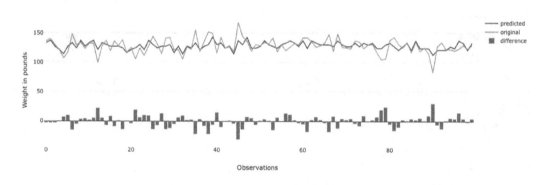

Figure 7.8 – Predicted versus original weights

We have learned about how a simple regression model is created using Redshift ML. Now let's learn about the multi-input regression model.

Creating multi-input regression models

In this exercise, you will learn how to build a regression model using multiple input variables in Amazon Redshift ML.

In this use case, we will use a dataset containing the sales history of online sporting events. A sporting event management company wants to review the data for the latest football and baseball seasons to figure out which games underperformed for revenue, and what the revenue projections look like for the season.

Your task is to build and train a model to predict revenue for upcoming events in order to proactively take action to increase ticket sales to ensure revenue numbers meet the company's targets.

After successfully connecting to Redshift as an admin or database developer, load data into Amazon Redshift.

Navigate to **Redshift query editor v2** and connect to the **Serverless** endpoint and the **dev** database.

Use the same schema and **query editor** page you created for the previous exercise.

Create your input table and load the data using the following SQL commands:

```
CREATE TABLE chapter7_RegressionModel.sporting_event_ticket_info (
ticket_id double precision ,
event_id bigint,
sport character varying(500),
event_date_time timestamp without time zone,
home_team character varying(500),
away_team character varying(500),
location character varying(500),
city character varying(500),
seat_level bigint,
seat_section bigint,
seat_row character varying(500),
seat bigint ENCODE az64,
list_ticket_price double precision,
final_ticket_price double precision ,
ticketholder character varying(500)
)
DISTSTYLE AUTO;
COPY chapter7_RegressionModel.sporting_event_ticket_info
FROM 's3://packt-serverless-ml-redshift/chapter07/ticket_info'
IAM_ROLE default
FORMAT AS CSV DELIMITER ',' QUOTE '"'
REGION AS 'eu-west-1';
```

Let's analyze our dataset and get a historical trend of ticket sales over the last few months:

```
Select extract(month from event_date_time) as month,
sum(cast (final_ticket_price as decimal(8,2))) as ticket_revenue
From chapter7_RegressionModel.sporting_event_ticket_info
where event_date_time < '2019-10-27'
group by 1
order by 1;
```

The output is as follows:

☑ month	☰	ticket_revenue
☑ 4		9452100.91
☑ 5		8745572.16
☑ 6		10319796.7
☑ 7		8729327.06
☑ 8		8457197.08
☑ 9		22048196.31
☑ 10		15344676.75

Figure 7.9 – Ticket revenue by month

We can see that sales are spiky and fall off dramatically in months **7** and **8**. Let's create a model so we can predict teams that will have lower ticket revenue. Before creating our model, we need to split the dataset into training, validation, and testing datasets, respectively.

1. Execute the following code in Query Editor v2 to create the `training` table:

```
CREATE TABLE chapter7_RegressionModel.sporting_event_ticket_
info_training (
    ticket_id double precision ,
    event_id bigint,
    sport character varying(500),
    event_date_time timestamp without time zone,
    home_team character varying(500),
    away_team character varying(500),
    location character varying(500),
    city character varying(500),
    seat_level bigint,
    seat_section bigint,
    seat_row character varying(500),
```

```
      seat bigint ENCODE az64,
      list_ticket_price double precision,
      final_ticket_price double precision ,
      ticketholder character varying(500)
  )
  DISTSTYLE AUTO;
```

2. The next step is to insert ~70% of the data into the `training` table:

```
--insert ~70% of data into training_set

insert into    chapter7_RegressionModel.sporting_event_ticket_
info_training
( ticket_id ,event_id ,sport , event_date_time,  home_team ,
away_team , location , city , seat_level, seat_section,
    seat_row ,  seat, list_ticket_price, final_ticket_price,
ticketholder )
 select
 ticket_id ,event_id ,sport , event_date_time,  home_team ,
away_team , location , city , seat_level, seat_section,
    seat_row ,  seat, list_ticket_price, final_ticket_price,
ticketholder
 from chapter7_RegressionModel.sporting_event_ticket_info
 where event_date_time < '2019-10-20';
```

3. Next, you will create the `validation` table:

```
CREATE TABLE chapter7_RegressionModel.sporting_event_ticket_
info_validation (
    ticket_id double precision ,
    event_id bigint,
    sport character varying(500),
    event_date_time timestamp without time zone,
    home_team character varying(500),
    away_team character varying(500),
    location character varying(500),
    city character varying(500),
    seat_level bigint,
    seat_section bigint,
    seat_row character varying(500),
    seat bigint ENCODE az64,
    list_ticket_price double precision,
    final_ticket_price double precision ,
    ticketholder character varying(500)
)
DISTSTYLE AUTO;
```

4. Next, insert ~10% of the data into the `validation` table:

```
insert into  chapter7_RegressionModel.sporting_event_ticket_
info_validation
( ticket_id ,event_id ,sport , event_date_time,  home_team ,
away_team , location , city , seat_level, seat_section,
    seat_row ,  seat, list_ticket_price, final_ticket_price,
ticketholder )
 select
 ticket_id ,event_id ,sport , event_date_time,  home_team ,
away_team , location , city , seat_level, seat_section,
    seat_row ,  seat, list_ticket_price, final_ticket_price,
ticketholder
 from chapter7_RegressionModel.sporting_event_ticket_info
 where event_date_time between '2019-10-20' and '2019-10-22' ;
```

5. Finally, create the `testing` table:

```
CREATE TABLE chapter7_RegressionModel.sporting_event_ticket_
info_testing (
    ticket_id double precision ,
    event_id bigint,
    sport character varying(500),
    event_date_time timestamp without time zone,
    home_team character varying(500),
    away_team character varying(500),
    location character varying(500),
    city character varying(500),
    seat_level bigint,
    seat_section bigint,
    seat_row character varying(500),
    seat bigint ENCODE az64,
    list_ticket_price double precision,
    final_ticket_price double precision ,
    ticketholder character varying(500)
)
DISTSTYLE AUTO;
```

6. Next, insert ~20% of the data into the `testing` table:

```
insert into   chapter7_RegressionModel.sporting_event_ticket_
info_testing
( ticket_id ,event_id ,sport , event_date_time,  home_team ,
away_team , location , city , seat_level, seat_section,
    seat_row ,  seat, list_ticket_price, final_ticket_price,
ticketholder )
select
```

```
    ticket_id ,event_id ,sport , event_date_time,  home_team ,
away_team , location , city , seat_level, seat_section,
    seat_row ,  seat, list_ticket_price, final_ticket_price,
ticketholder
 from chapter7_RegressionModel.sporting_event_ticket_info
 where event_date_time >  '2019-10-22'
 ;
```

We have prepared the dataset to train and test the ML model; now let's create a regression model using the Linear Learner algorithm.

Linear Learner algorithm

As we saw in *Chapter 6*, you can use the Linear Learner model type to solve classification or regression problems. This is a supervised learning algorithm. For regression problems, we are trying to predict a numerical outcome and, in this exercise, we will be using multiple inputs; SageMaker will choose the best modes based on continuous objectives using MSE.

We provide a training set with data that contains our inputs or observations about the data and the label, which represents the value we want to predict. Our goal is to accurately predict future ticket sales.

We set MODEL_TYPE as LINEAR_LEARNER, PROBLEM_TYPE as regression, and OBJECTIVE as mse. We leave out other options as default.

Execute this code in Query Editor v2 to train the model. Be sure to replace the following S3 bucket using the bucket you created previously. You will need to input the S3 bucket you created previously to store the Redshift ML artifacts.

Run the following command to train the regression model:

```
CREATE MODEL chapter7_RegressionModel.predict_ticket_price_linlearn
from
chapter7_RegressionModel.sporting_event_ticket_info_training
TARGET final_ticket_price
FUNCTION predict_ticket_price_linlearn
IAM_ROLE default
MODEL_TYPE LINEAR_LEARNER
PROBLEM_TYPE regression
OBJECTIVE 'mse'
SETTINGS (s3_bucket '<<your-S3-Bucket>>',
s3_garbage_collect off,
max_runtime 9600);
```

Once the model state is **READY**, you are ready to proceed. To check the status of the model, run the following command:

```
SHOW MODEL chapter7_RegressionModel.predict_ticket_price_linlearn;
```

Note the MSE score you see; it will be similar to the output in *Figure 7.10*:

Key	Value
Model Name	predict_ticket_price_linlearn
Schema Name	chapter7_regressionmodel
Owner	IAMR:Admin
Creation Time	Wed, 15.02.2023 13:32:14
Model State	READY
validation:mse	681.905884
Estimated Cost	5.903461
TRAINING DATA:	
Query	SELECT *
	FROM "CHAPTER7_REGRESSIONMODEL"."SPORTING_EVENT_TICKET_INFO_TRAINING"
Target Column	FINAL_TICKET_PRICE
PARAMETERS:	
Model Type	linear_learner
Problem Type	Regression
Objective	MSE
AutoML Job Name	redshiftml-20230215133214851755
Function Name	predict_ticket_price_linlearn

Figure 7.10 – SHOW MODEL output

We have now created the ML model; let's validate its performance.

Understanding model evaluation

You measure the model performance of regression problems through the MSE and/or RMSE. This is the distance between the predicted numeric target and the actual numeric answer, also known as **ground truth**. In our SHOW MODEL output, we see the MSE. We can also calculate this ourselves by squaring the differences between the actual and predicted values and then finding the average. Then, take the square root of MSE to get the RMSE. The lower the MSE and RMSE scores, the better.

As we see from the SHOW MODEL output, our MSE score is over 681 – let's check this and the RMSE score against our validation by running the following SQL command:

```
SELECT
       ROUND(AVG(POWER(( actual_price_revenue - predicted_price_revenue
),2)),2) mse
```

```
     , ROUND(SQRT(AVG(POWER(( actual_price_revenue - predicted_price_
revenue ),2)))),2) rmse
FROM
     (select home_team, chapter7_RegressionModel.predict_ticket_price_
linlearn (ticket_id, event_id, sport, event_date_time, home_team,
away_team,
Location, city, seat_level, seat_section, seat_row, seat,
list_ticket_price ,ticketholder ) as predicted_price_revenue,
 final_ticket_price  as actual_price_revenue
From chapter7_RegressionModel.sporting_event_ticket_info_validation
     );
```

This is the output of the query:

Figure 7.11 – MSE and RMSE values

While the MSE scores seem a little high, we can also run a validation query to check our accuracy rates. You will notice in the following query that it uses the function that was generated by our CREATE MODEL command to get the predicted price revenue for us to compare to the actual price revenue:

```
Select home_team,
sum(cast(chapter7_RegressionModel.predict_ticket_price_linlearn
(ticket_id, event_id, sport,
event_date_time, home_team, away_team,
Location, city, seat_level, seat_section, seat_row, seat,
list_ticket_price ,ticketholder ) as decimal(8,2) )) as predicted_
price_revenue,
sum(cast (final_ticket_price as decimal(8,2))) as actual_price_
revenue,
(predicted_price_revenue - actual_price_revenue) as diff,
abs((predicted_price_revenue - actual_price_revenue)/actual_price_
revenue) * 100  as pct_diff
From chapter7_RegressionModel.sporting_event_ticket_info_validation
group by 1
order by 5 desc ;
```

This is the output of the query:

home_team	predicted_price_revenue	actual_price_revenue	diff	pct_diff
☐ Arizona Piranga	1547759.42	999293.1	548466.32	54.88
☐ Minnesota Miners	729309.15	478958.86	250350.29	52.26
☐ New Orleans Chrub	1095761.42	744172.41	351589.01	47.24
☐ Philadelphia Yellowja...	418519.98	691901.1	-273381.12	39.51
☐ San Francisco Gold	718069.13	522790.97	195278.16	37.35
☐ Los Angeles Angles	129775.16	95822.37	33952.79	35.43
☐ Carolina Cheetah	274939.84	228438.44	46501.4	20.35
☐ St. Louis Nobels	63127.87	56264.75	6863.12	12.19
☐ Seattle Ospreys	791926.67	891482.93	-99556.26	11.16
☐ Los Angeles Fordhams	347899.25	314726.17	33173.08	10.54
☐ Kansas City Regals	86508.24	96022.64	-9514.4	9.9

Figure 7.12 – Predicted price versus actual price

Looking at the results, the model is not performing as well as we would like. You can run the validation query against the training data and see that the model is not performing very well on the training data either – this is called **underfitting**.

One solution would be to add more features, but we have already used all the available features.

Let's try running the model again, but this time, we will use the auto option and let SageMaker pick the algorithm:

```
CREATE MODEL Chapter7_RegressionModel.predict_ticket_price_auto
from
chapter7_RegressionModel.sporting_event_ticket_info_training
TARGET final_ticket_price
FUNCTION predict_ticket_price_auto
IAM_ROLE default
PROBLEM_TYPE regression
OBJECTIVE 'mse'
SETTINGS (s3_bucket '<<your-S3-bucket>>',
s3_garbage_collect off,
max_runtime 9600);
```

After letting the model train for some time, check the status of the model as follows:

```
SHOW MODEL Chapter7_RegressionModel.predict_ticket_price_auto;
```

This is how it appears:

Key	Value
Model Name	predict_ticket_price_auto
Schema Name	chapter7_regressionmodel
Owner	IAMR:Admin
Creation Time	Wed, 15.02.2023 17:46:38
Model State	READY
validation:mse	0.800670
Estimated Cost	8.401197
TRAINING DATA:	
Query	SELECT *
	FROM "CHAPTER7_REGRESSIONMODEL"."SPORTING_EVEN...
Target Column	FINAL_TICKET_PRICE
PARAMETERS:	
Model Type	xgboost
Problem Type	Regression
Objective	MSE
AutoML Job Name	redshiftml-20230215174638030561
Function Name	predict_ticket_price_auto
Function Parameters	ticket_id event_id sport event_date_time home_team away_tea...
Function Parameter T...	float8 int8 varchar timestamp varchar varchar varchar varchar in...
IAM Role	default-aws-iam-role

Figure 7.13 – SHOW MODEL output

From the preceding figure, we see that two things stand out:

- The MSE score is much better
- Amazon SageMaker chose to use the XGBoost algorithm

We can check the MSE and RMSE scores for our new model using our validation dataset as follows:

```
SELECT
     ROUND(AVG(POWER(( actual_price_revenue - predicted_price_revenue
),2)),2) mse
     , ROUND(SQRT(AVG(POWER(( actual_price_revenue - predicted_price_
revenue ),2))),2) rmse
FROM
     (select home_team, chapter7_RegressionModel.predict_ticket_price_
auto (ticket_id, event_id, sport, event_date_time, home_team, away_
team,
Location, city, seat_level, seat_section, seat_row, seat,
list_ticket_price ,ticketholder ) as predicted_price_revenue,
 final_ticket_price  as actual_price_revenue
From chapter7_RegressionModel.sporting_event_ticket_info_validation
     );
```

This is the output:

☐ mse	rmse
☐ 3.57999999999999996	1.89

Figure 7.14 – MSE and RMSE scores

These MSE and RMSE values show that we have a good model.

Let's run a validation query using the predict_ticket_price_auto function from the new model:

```
Select home_team,
sum(cast(chapter7_RegressionModel.predict_ticket_price_auto (ticket_
id, event_id, sport, event_date_time, home_team, away_team,
Location, city, seat_level, seat_section, seat_row, seat,
list_ticket_price ,ticketholder ) as decimal(8,2) )) as predicted_
price_revenue,
sum(cast (final_ticket_price as decimal(8,2))) as actual_price_
revenue,
(predicted_price_revenue - actual_price_revenue) as diff,
((predicted_price_revenue - actual_price_revenue)/actual_price_
revenue) * 100  as pct_diff
From chapter7_RegressionModel.sporting_event_ticket_info_validation
group by 1
order by 5 desc;
```

The following is the output for this query:

home_team	predicted_price_revenue	actual_price_revenue	diff	pct_diff
New Orleans Chrub	757229.54	744172.41	13057.13	1.75
Atlanta Kestrels	539584.2	534735.19	4849.01	0.9
Los Angeles Angles	96690.08	95822.37	867.71	0.9
St. Louis Nobels	56754.8	56264.75	490.05	0.87
Dallas Horsemen	348109.78	345391.58	2718.2	0.78
Seattle Oceans	140937.7	139886.02	1051.68	0.75
Carolina Cheetah	229954.53	228438.44	1516.09	0.66
New York Behemoths	636323.58	632688.58	3635	0.57
Arizona Silverbacks	120331.11	119670.29	660.82	0.55
Kansas City Regals	96379.57	96022.64	356.93	0.37
Minnesota Pair	106673.41	106376.56	296.85	0.27
New York Janke	133939.1	133591.91	347.19	0.25
San Diego Pastors	93606.3	93448.18	158.12	0.16
New York Meets	97857.3	97812.15	45.15	0.04
Washington Citizens	83148.62	83115.21	33.41	0.04
Texas Tejanos	119226.38	119186.13	40.25	0.03
San Francisco Gold	522960.49	522790.97	169.52	0.03

Figure 7.15 – Predicted price versus actual price

You can see we have much better results when comparing the differences between the actual and predicted ticket price revenue. We will use this model to do our prediction queries.

Run the following query to see which inputs contributed most to the model prediction:

```
set search_path to chapter7_regressionmodel;

select json_table.report.explanations.kernel_shap.label0.global_shap_
values from
  (select explain_model('predict_ticket_price_auto') as report) as
json_table
```

To make the result set easier to read, right-click on the result set and choose **Copy rows**. You can then paste that into the editor as shown in *Figure 7.16*:

```
{"away_team":0.062348301666640498,"city":0.013471148633644594,"event_date_time":0.08475716160351603,
"event_id":0.5273868896552291,"home_team":0.10166936254900883,"list_ticket_price":35.65163475556435,
"location":0.01303543586712852,"seat":0.17608651044634714,"seat_level":0.02886946029532519,
"seat_row":0.09813160435858882,"seat_section":12.258673814862848,"sport":0.008466236038277474,
"ticket_id":0.4721946570286829,"ticketholder":0.015446279672691717}
```

Figure 7.16 – Model explainability report

This shows that `list_ticket_price` contributed the most weight and `sport` contributed the least weight.

We have validated the model with a validation dataset, checked the MSE values, and determined feature importance. Now let's run the prediction query against test data.

Prediction query

Now that we have our model and have done validation, we can run our prediction query against our test dataset to determine which teams and events will need a proactive approach to increase ticket sales. Let's check for teams with a predicted revenue of less than 200K:

```
select t1.home_team, predicted_price_revenue
from
(Select home_team,
sum(cast(chapter7_RegressionModel.predict_ticket_price_auto (ticket_
id, event_id, sport, event_date_time, home_team, away_team,
Location, city, seat_level, seat_section, seat_row, seat,
list_ticket_price ,ticketholder ) as decimal (8,2) ) ) as predicted_
price_revenue
From chapter7_RegressionModel.sporting_event_ticket_info_testing
group by 1) t1
where predicted_price_revenue < 200000;
```

This is the result:

home_team	predicted_price_revenue
St. Louis Nobels	88861.85
Kansas City Regals	58583.65
Colorado Stones	151493.93
Washington Citizens	79939.99
Los Angeles Angles	115151.41
Oakland Clubs	94642.13
Arizona Silverbacks	93533.08
Detroit Lynx	88446.93
Chicago Snowy Sox	69344.39
Pittsburgh Alleghenys	86476.07
Tampa Bay Sting	55768.95
Chicago Chaps	98611.8
Baltimore Oreos	144523.03
San Francisco Goliaths	123145.88
Cincinnati Maroon	102373.7
New York Janke	132508.22

Figure 7.17 – Predicted price against the test dataset

There are 16 teams that are predicted to have reduced ticket revenue. You can share this information with your marketing teams to create a focused strategy to ensure ticket revenues can remain on track.

Summary

In this chapter, we discussed regression models in detail and saw how to create single-input and multi-input regression models. We learned how easy it is to predict a numeric value. We also learned how to validate regression models, take actions to improve our model's accuracy, and do prediction queries with our regression models. We walked through options for using XGBoost, Linear Learner and `auto` options to train your model.

We also saw how we can check and validate the MSE score from the SHOW MODEL output using SQL commands in Redshift.

In the next chapter, we will show you how to create unsupervised models using the K-means algorithm.

8

Building Unsupervised Models with K-Means Clustering

So far, we have learned about building **machine learning** (**ML**) models where data is supplied with labels. In this chapter, we will learn about building ML models on a dataset without any labels by using the **K-means clustering algorithm**. Unlike **supervised models**, where predictions are made at the observation level, K-means clustering groups observations into clusters where they share a commonality – for example, similar demographics or reading habits.

This chapter will provide detailed examples of business problems that can be solved with these modeling techniques. By the end of this chapter, you will be in a position to identify a business problem that an **unsupervised modeling technique** can be applied to. You will also learn how to build, train, and evaluate K-means model performance.

In this chapter, we will cover the following main topics:

- Grouping data through cluster analysis

- Creating a K-means ML model

- Evaluating the results of K-means clustering

Technical requirements

This chapter requires a web browser and access to the following:

- An AWS account

- An Amazon Redshift Serverless endpoint

- Amazon Redshift Query Editor v2

- Complete the *Getting started with Amazon Redshift Serverless* section in *Chapter 1*

You can find the code used in this chapter here: `https://github.com/PacktPublishing/Serverless-Machine-Learning-with-Amazon-Redshift/blob/main/CodeFiles/chapter8/chapter8.sql`.

Grouping data through cluster analysis

So far, we have explored datasets that contained input and target variables, and we trained a model with a set of input variables and a target variable. This is called supervised learning. However, how do you address a dataset that does not contain a label to supervise the training? **Amazon Redshift ML** supports unsupervised learning using the cluster analysis method, also known as the K-means algorithm. In **cluster analysis**, the ML algorithm automatically discovers the grouping of data points. For example, if you have a population of 1,000 people, a clustering algorithm can group them based on height, weight, or age.

Unlike supervised learning, where an ML model predicts an outcome based on a label, unsupervised models use unlabeled data. One type of unsupervised learning is clustering, where unlabeled data is grouped based on its similarity or differences. From a dataset with demographic information about individuals, you can create clusters based on young, adult, and elderly populations, underweight, normal weight, and overweight populations, and so on. These groups are calculated based on values – for example, if two people are young, then they are grouped together. These groups are called **clusters**. In the following diagram, you can see that input variables (**Age**, **Height**, and **Weight**) are grouped into **young**, **adult**, and **elder**:

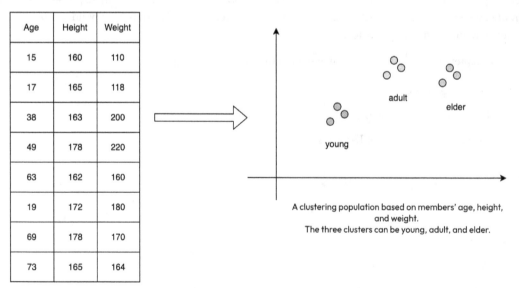

Figure 8.1 – A simple cluster example

In the preceding diagram, each individual data point is placed in a cluster, based on the distance from the center of the cluster, called the **centroid**. The distance from the centroid for each data point is calculated using the **Euclidean distance formula**. Data points that are closest to a given centroid have similarities and belong to the same group. In real-world situations, it is very common to find data points with overlapping clusters and too many of them. When you encounter too many clusters, then it is a challenge to identify the right number of clusters for your dataset.

Common use cases for a K-means cluster include the following:

- **E-commerce**: Grouping customers by purchase history
- **Healthcare**: Detecting patterns of diseases
- **Finance**: Grouping purchases into abnormal versus normal

Next, we will show you one of the common methods to help you determine how many clusters you should use.

Determining the optimal number of clusters

One popular method that is frequently adopted is the **Elbow method**. The idea of the Elbow method is to run K-means algorithms with different values of K – for example, from 1 cluster all the way to 10 – and for each value of K, calculate the sum of squared errors. Then, plot a chart of the **sum of squared deviation** (**SSD**) values. SSD is the sum of the squared difference and is used to measure variance. If the line chart looks like an arm, then the *elbow* on the arm is the value of K that is the best among the various K values. The method behind this approach is that SSD usually tends to decrease as the value of K is increased, and the goal of the evaluation method is also to aim for lower SSD or **mean squared deviation** (**MSD**) values. The elbow represents a starting point, where SSD starts to have diminishing returns when the K value increases.

In the following chart, you can see that the MSD value, when charted over different K values, represents an arm, and the *elbow* is at value **6**. After **6**, there is no significant decrease in the MSD value, so we can pick 6 as the best cluster value in the following scenario:

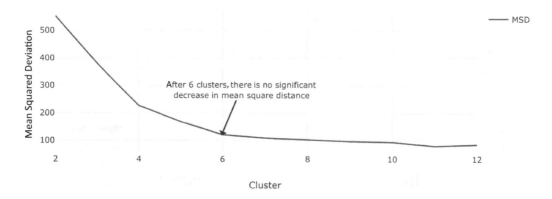

Figure 8.2 – MSD values when charted over different K values

Next, let's see how we can create a K-means clustering model with Amazon Redshift ML.

Creating a K-means ML model

In this section, we will walk through the process with the help of a use case. In this use case, assume you are a data analyst for an e-commerce company specializing in home improvement goods. You have been tasked with classifying economic segments in different regions, based on income, so that you can better target customers, based on various factors, such as median home value. We will use this dataset from Kaggle: `https://www.kaggle.com/datasets/camnugent/california-housing-prices`.

From this dataset, you will use the `median_income`, `latitude`, and `longitude` attributes so that you can create clusters based on `location` and `income`.

The syntax to create a K-means model is slightly different from what you will have used up to this point, so let's dive into that.

Creating a model syntax overview for K-means clustering

Here is the basic syntax to create a K-means model:

```
CREATE model model_name
FROM (Select_statement)
FUNCTION  function_name
IAM_ROLE default
AUTO OFF
MODEL_TYPE KMEANS
PREPROCESSORS (StandardScaler', 'MinMax', 'NumericPassthrough')
HYPERPARAMETERS DEFAULT EXCEPT (K '2')
SETTINGS (S3_BUCKET 'bucket name');
```

A couple of key things to note in the preceding code snippet are the lines in bold, as they are required when creating K-means models:

- `AUTO OFF`: This must be turned off, since Amazon SageMaker Autopilot is not used for K-means

- `MODEL_TYPE KMEANS`: You must set `MODEL_TYPE`, as there is no auto-discovery for K-means

- `HYPERPARAMETERS DEFAULT EXCEPT (K '2')`: This tells SageMaker how many clusters to create in this model

Also, note that there are three optional preprocessors available with K-means. We will explore that in more detail when we create the model.

You can refer to this link for more details on the K-means parameters available: `https://docs.aws.amazon.com/redshift/latest/dg/r_create_model_use_cases.html#r_k-means-create-model-parameters`.

Now, we will load our dataset in preparation for creating our model.

Uploading and analyzing the data

For this use case, we will use a file that contains housing price information and summary stats, based on census data.

> **Note**
> Data is stored in the following S3 location: `s3://packt-serverless-ml-redshift/chapter08/housinghousing_prices.csv`.

After successfully connecting to Redshift as an admin or database developer, load data into Amazon Redshift and follow the steps outlined here.

1. Navigate to Redshift **query editor v2**, connect to the **Serverless: default** endpoint, and then connect to the **dev** database.

Figure 8.3 – Connecting via Redshift query editor v2

2. Execute the following steps to create the schema and customer table and load the data:

```
create schema chapter8_kmeans_clustering;

create table chapter8_kmeans_clustering.housing_prices (
     longitude decimal(10,2),
     latitude decimal(10,2),
     housing_median_age integer,
     total_rooms integer,
     total_bedrooms integer,
     population integer,
     households integer,
     median_income decimal(10,6),
     median_house_value integer,
     ocean_proximity character varying (50)
)
diststyle auto;

copy chapter8_kmeans_clustering.housing_prices from 's3://packt-
serverless-ml-redshift/chapter08/kmeans/housing_prices.csv'
iam_role default format as csv delimiter ',' quote '"'
ignoreheader 1 region as 'eu-west-1';
```

This dataset contains 2,064,020,640 records. We will use `longitude`, `latitude`, and `median_income` in our model.

3. Run the following query to examine some sample data:

```
select * from chapter8_kmeans_clustering.housing_prices
limit 10;
```

You should get the following result:

longitude	latitude	housing_median_age	total_rooms	total_bedrooms	population	households	median_income
-122.23	37.88	41	880	129	322	126	8.3252
-122.22	37.86	21	7099	1106	2401	1138	8.3014
-122.24	37.85	52	1467	190	496	177	7.2574
-122.25	37.85	52	1274	235	558	219	5.6431
-122.25	37.85	52	1627	280	565	259	3.8462
-122.25	37.85	52	919	213	413	193	4.0368
-122.25	37.84	52	2535	489	1094	514	3.6591
-122.25	37.84	52	3104	687	1157	647	3.12
-122.26	37.84	42	2555	665	1206	595	2.0804
-122.25	37.84	52	3549	707	1551	714	3.6912

Figure 8.4 – Housing prices data

Now that the data is loaded, we are ready to create the model.

Creating the K-means model

Let's create our model and cluster based on `median_income`, `longitude`, and `latitude`.

We will create a few models and then use the elbow method to determine the optimal number of clusters.

To begin with, let's create our first model with two clusters using the following SQL. You can then experiment by creating different models by changing the K value, and then you can learn how the MSD value diminishes over different K values.

Creating two clusters with a K value of 2

Let's run the following SQL in Query Editor v2 to create a model with two clusters:

```
create model chapter8_kmeans_clustering.housing_segments_k2
from(select
            median_income,
            latitude,
            longitude
from chapter8_kmeans_clustering.housing_prices)
function  get_housing_segment_k2
iam_role default
auto off
model_type kmeans
preprocessors '[
      {
        "columnset": [ "median_income", "latitude","longitude" ],
        "transformers": [ "standardscaler" ]
      }
    ]'
```

```
hyperparameters default except (k '2')

settings (s3_bucket '<your s3 bucket>');
```

You can see in this model that we supply values for the `preprocessors` parameter. We chose to do this because K-means is sensitive to scale, so we can normalize with the `standardscaler` transformer. `standardscalar` moves the mean and scale to unit variance.

The `hyperparameters` parameter is where we specify `(K '2')` to create two clusters. Remember to add your S3 bucket, where the created model artifacts are stored. You will find the model artifacts in `s3: s3://<your-s3-bucket>/redshift-ml/housing_segments_k2/`. Redshift ML will automatically append `'redshift-ml'/'your model name'` to your S3 bucket. Now, check the status of the model, using the `SHOW MODEL` command in Query Editor v2:

```
SHOW MODEL chapter8_kmeans_clustering.housing_segments_k2;
```

You will see the following output:

Key	Value
Model Name	housing_segments_k2
Schema Name	chapter8_kmeans_clustering
Owner	demo
Creation Time	Tue, 24.01.2023 03:55:05
Model State	READY
train:msd	1.088200
train:progress	100.000000
train:throughput	249274.187500
Estimated Cost	0.007156
TRAINING DATA:	
Query	SELECT MEDIAN_INCOME, LATITUDE, LONGITUDE
	FROM CHAPTER8_KMEANS_CLUSTERING.HOUSING_PRICES
PARAMETERS:	
Model Type	kmeans

Figure 8.5 – Two clusters

The key things to note are **Model State**, which indicates that the model is ready, and **train:msd**, which is the objective metric. This represents the mean squared distances between each record in our input dataset and the closest center of the model. The **MSD** value is **1.088200**, which is a good score.

Let's run a query to get the number of data points in each cluster:

```
select chapter8_kmeans_clustering.get_housing_segment_k2 (median_
income, latitude, longitude) as cluster, count(*) as count from FROM
chapter8_kmeans_clustering.housing_prices group byGROUP BY 1 order
byORDER BY 1;
```

The output is as follows:

☐	cluster	count
☐	0	8719
☐	1	11921

Figure 8.6 – The data points

Clusters are numbered from 0 to n. Our first cluster has **8719** data points, and the second cluster has **11921** data points.

In our use case, we want to further segment our customers. Let's create a few more models with different numbers of clusters. We can then evaluate all the SSD values and apply the Elbow method to help us choose the optimal number of clusters to use for our analysis.

Creating three clusters with a K value of 3

Let's run the following SQL in Query Editor v2 to create a model with three clusters:

```
CREATE model chapter8_kmeans_clustering.housing_segments_k3
FROM(Select
            median_income,
            latitude,
            longitude
From chapter8_kmeans_clustering.housing_prices)
FUNCTION  get_housing_segment_k3
IAM_ROLE default
AUTO OFF
MODEL_TYPE KMEANS
PREPROCESSORS '[
    {
        "ColumnSet": [ "median_income", "latitude","longitude" ],
        "Transformers": [ "StandardScaler" ]
    }
    ]'
HYPERPARAMETERS DEFAULT EXCEPT (K '3')
SETTINGS (S3_BUCKET '<your s3 bucket>');
```

Creating the remaining models with clusters 4, 5, and 6

Repeat the preceding code 3 more times to create models with 4, 5, and 6 clusters, respectively. You will find the code at `https://github.com/PacktPublishing/Serverless-Machine-Learning-with-Amazon-Redshift/blob/main/CodeFiles/chapter8/chapter8.sql`.

It will take ~15 minutes for all the models to finish training. Then, run the SHOW MODEL command, including the one for the model where K = 2, as shown here:

```
SHOW MODEL chapter8_kmeans_clustering.housing_segments_k2;
SHOW MODEL chapter8_kmeans_clustering.housing_segments_k3;
SHOW MODEL chapter8_kmeans_clustering.housing_segments_k4;
SHOW MODEL chapter8_kmeans_clustering.housing_segments_k5;
SHOW MODEL chapter8_kmeans_clustering.housing_segments_k6;
```

Now, let's find the elbow!

Gathering inputs to chart the elbow

Now, from the output of each SHOW MODEL command, note the value for test:msd and build a Select statement, as shown in the following code snippet. Change the value for MSD using the test:mds value for each model.

As an example, we will use the value 1.088200, which we saw earlier for train:msd, for the model with two clusters.

Our other output from train:mds from the SHOW MODEL output is as follows:

- Two clusters: train:msd – 1.088200

- Three clusters: train:msd – 0.775993

- Four clusters: train:msd – 0.532355

- Five clusters: train:msd – 0.437294

- Six clusters: train:msd – 0.373781

Note that your numbers may be slightly different:

```
Select 2 as K, 1.088200  as MSD
Union
Select 3 as K,  0.775993 as MSD
Union
Select 4 as K,  0.532355 as MSD
Union
Select 5 as K,  0.437294 as MSD
Union
Select 6 as K,  0.373781 as MSD;
```

Run the preceding SQL command in Query Editor v2.

By observing the output, we can see that the MSD value is highest for two clusters and gradually decreases as the number of clusters increases:

k	msd
2	1.0882
3	0.775993
4	0.532355
5	0.437294
6	0.373781

Figure 8.7 – msd

In the **Result** window, click on the **Chart** option, as shown here:

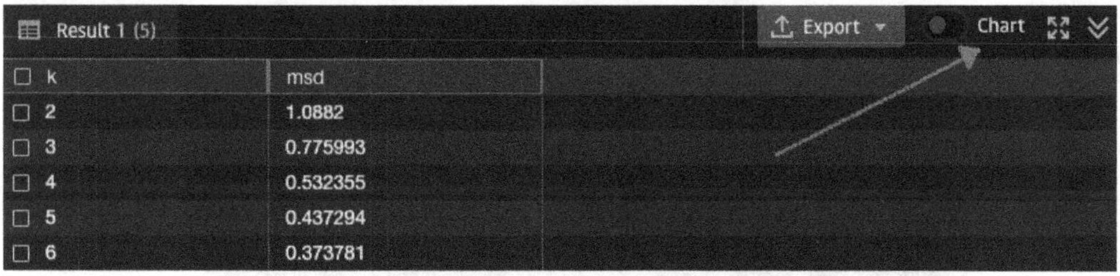

Result 1 (5)		Export ▾ ● Chart ⤢ ⌄
k	msd	
2	1.0882	
3	0.775993	
4	0.532355	
5	0.437294	
6	0.373781	

Figure 8.8 – Creating a chart

By choosing `k` as the X value and `msd` as the Y value, you will get the following output:

Figure 8.9 – The elbow method chart

From the chart, we can see that when MSD is charted over a line graph, an arm is formed, and the elbow is at **3**. This means that there is little difference in the MSD value with **4** clusters compared to **3** clusters . We can see that after **3**, the curve is very smooth, and the difference between the MSD value does not drastically change compared to the beginning of the line.

Let's see how data points are clustered when we use a function deployed for our model with three clusters:

```
select chapter8_kmeans_clustering.get_housing_segment_k3 (median_
income, latitude, longitude) as cluster, count(*) as count from
chapter8_kmeans_clustering.housing_prices group by 1 order by 1;
```

We can see the following output from Query Editor v2. The counts represent the number of data points assigned to each cluster:

cluster	count
0	9303
1	8425
2	2912

Figure 8.10 – Three clusters

We can also chart this by clicking on the **Chart** button and observing the cluster counts represented visually:

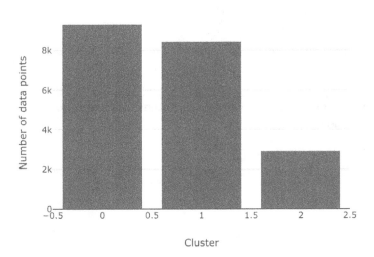

Figure 8.11 – The cluster data points

Now, let's see how we can use our model to help make business decisions based on the clusters.

Evaluating the results of the K-means clustering

Now that you have segmented your clusters with the K-means algorithm, you are ready to perform various analyses using the model you created.

Here is an example query you can run to get the average median house value by cluster:

```
select avg(median_house_value) as avg_median_house_value,
chapter8_kmeans_clustering
.get_housing_segment_k3(median_income, latitude, longitude) as cluster
from chapter8_kmeans_clustering
.housing_prices
group by 2
order by 1;
```

The output will look like this:

avg_meadian_hous...	cluster
178924	0
193582	1
334492	2

Figure 8.12 – Average median house values

You can also run a query to see whether higher median incomes correspond to the same clusters with higher home values. Run the following query:

```
select avg(median_income) as median_income,
chapter8_kmeans_clustering.get_housing_segment_k3(
    median_income, latitude, longitude) as cluster
from chapter8_kmeans_clustering.housing_prices
group by 2
order by 1;
```

The output will look like this:

median_income ≡	cluster
3.100833	0
3.685455	1
6.865946	2

Figure 8.13 – median_income

When we established our use case, we said this was for an e-commerce retailer specializing in home improvement products. Another way you could use this information is to create different marketing campaigns and tailor your product offerings, based on home values in a given cluster.

Summary

In this chapter, we discussed how to do unsupervised learning with the K-means algorithm.

You are now able to explain what the K-means algorithm is and what use cases it is appropriate for. Also, you can use Amazon Redshift ML to create a K-means model, determine the appropriate number of clusters, and draw conclusions by analyzing the clusters to help make business decisions.

In the next chapter, we will show you how to use the multi-layer perceptron algorithm to perform deep learning with Amazon Redshift ML.

Part 3:
Deploying Models with Redshift ML

Part 3 introduces you to more ways to leverage Amazon Redshift ML. You will learn about deep learning algorithms, how to train a customized model, and how you can use models trained outside of Amazon Redshift to run inference queries in your data warehouse.

This part closes with an introduction to time-series forecasting, how to use it with Amazon Redshift ML, and how you can optimize and easily re-train your models.

This part comprises the following chapters:

- *Chapter 9, Deep Learning with Redshift ML*
- *Chapter 10, Creating Custom ML Models with XGBoost*
- *Chapter 11, Bring Your Own Models for In-Database Inference*
- *Chapter 12, Time-Series Forecasting in Your Data Warehouse*
- *Chapter 13, Operationalizing and Optimizing Amazon Redshift ML Models*

9

Deep Learning with Redshift ML

We explored **supervised learning** in *Chapters 6* and *7* and **unsupervised learning** models in *Chapter 8*. In this chapter, we will explore **deep learning algorithms**, a **multilayer perceptron** (**MLP**), which is a **feedforward artificial neural network** (**ANN**), and understand how it handles data that is not linearly separable (which means the data points in your data cannot be separated by a clear line). This chapter will provide detailed steps on how to perform deep learning in Amazon Redshift ML using MLP. By the end of this chapter, you will be in a position to identify a business problem that can be solved using MLP and know how to create the model, evaluate the performance of the model, and run predictions.

In this chapter, we will go through the following main topics:

- Introduction to deep learning
- Business problem
- Uploading and analyzing the data
- Creating a multiclass classification model using MLP
- Running predictions

Technical requirements

This chapter requires a web browser and access to the following:

- AWS account
- Amazon Redshift Serverless endpoint
- Amazon Redshift Query Editor v2

You can find the code used in this chapter here: `https://github.com/PacktPublishing/`
`Serverless-Machine-Learning-with-Amazon-Redshift/blob/main/CodeFiles/`
`chapter9/chapter9.sql`.

Introduction to deep learning

Deep learning is a type of **artificial intelligence (AI)** that uses algorithms to analyze and learn data to draw output similar to the way humans do. Deep learning can leverage both supervised and unsupervised learning using **artificial neural networks (ANNs)**. In deep learning, a set of outputs is generated from the input layers using a feedforward ANN called an MLP. The MLP utilizes backpropagation to feed the errors from the outputs back into the layers to compute one layer at a time and iterates until the model has learned the patterns and relationships in the input data to arrive at a specific output.

Feature learning is a set of techniques where the machine uses raw data to derive the characteristics of a class in the data to derive a specific task at hand. Deep learning models use feature learning efficiently to learn complex, redundant, and variable input data and classify the specified task. Thus, it eliminates the need for manual feature engineering for designing and selecting the input features. Deep learning is very useful when your datasets cannot be separated by a straight line, known as non-linear data.

For example, in classifying financial transactions as fraudulent or legitimate, there may not be a clear linear boundary between the two classes of data. In such cases, deep learning models can learn these variable and complex non-linear relationships between the features of the input data and thus improve the accuracy of the target classification.

When working on classification problems, an easy way to determine whether your dataset is linearly separated is to draw a scatter plot for classes and see whether two classes can be separated by a line or not. In the following diagram, the left-hand chart shows that two classes are linearly separated and the right-hand chart shows that they are not:

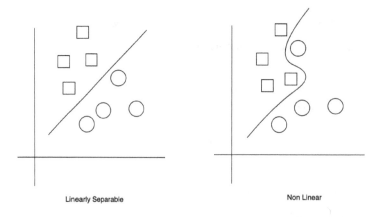

Figure 9.1 – Linear versus non-linear datasets

You can create models in Redshift ML when your dataset cannot be linearly separated by using the MLP algorithm. Common use cases where MLP algorithms are useful are as follows:

- Speech recognition
- Image recognition
- Machine translation

This chapter will show you how to build deep learning models in Amazon Redshift ML using the MLP algorithm.

Business problem

We will use a wall-following robot navigation dataset to build a machine learning model using the MLP algorithm. The robot is equipped with ultrasound sensors and data is collected as the robot navigates through the room in a clockwise direction. The goal here is to guide the robot to follow the wall by giving simple directions such as *Move-Forward*, *Slight-Right-Turn*, *Sharp-Right-Turn*, and *Slight-Left-Turn*.

Since there are classes to predict for a given set of sensor readings, this is going to be a multiclass problem. We will use MLP to correctly guide the robot to follow the wall. (This data is taken from `https://archive.ics.uci.edu/ml/datasets/Wall-Following+Robot+Navigation+Data` and is attributed to Ananda Freire, Marcus Veloso, and Guilherme Barreto (2019). UCI Machine Learning Repository [`http://archive.ics.uci.edu/ml`]. Irvine, CA: University of California, School of Information and Computer Science.)

Please follow the detailed document on the page to gain more understanding of the use case.

Now, you will upload the data, analyze it, and prepare for training the model.

Uploading and analyzing the data

We have sensor readings data stored in the following S3 location:

`s3://packt-serverless-ml-redshift/chapter09/`

After successfully connecting to Redshift as an admin or database developer, load data into Amazon Redshift:

1.   Navigate to **Redshift query editor v2** and connect to **Serverless: workgroup2** and then to the **dev** database:

Figure 9.2 – Connect to the dev database

2.   Execute the following steps to create the schema and customer table, and load the data:

```
create schema chapter9_deeplearning;

create table chapter9_deeplearning.robot_navigation (
     id bigint identity(0,1),
us1 float, us2 float, us3 float, us4 float, us5 float, us6
float, us7 float, us8 float, us9 float,us10 float, us11 float,
us12 float, us13 float, us14 float, us15 float, us16 float,
us17 float, us18 float, us19 float, us20 float, us21 float, us22
float, us23 float, us24 float, direction varchar(256)
)
diststyle auto;

copy chapter9_deeplearning.robot_navigation from 's3://packt-
serverless-ml-redshift/chapter09/sensor_readings_24.data'
iam_role   default
format as csv
delimiter ','
quote '"'
region as 'eu-west-1'
;
```

3.   Run the following query to examine some sample data:

```
select * from
chapter9_deeplearning.robot_navigation
limit 10;
```

In *Figure 9.3*, we can see that our data has been loaded successfully:

Id	us1	us2	us3	us4	us5
1	0.438	0.498	3.625	3.645	5
2	0.438	0.498	3.625	3.648	5
3	0.438	0.498	3.625	3.629	5
4	0.437	0.501	3.625	3.626	5
5	0.438	0.498	3.626	3.629	5
6	0.439	0.498	3.626	3.629	5
7	0.44	5	3.627	3.628	5
8	0.444	5.021	3.631	3.634	5

Figure 9.3 – Sample output

From the preceding screenshot, we can see that there are several sensor readings. Run the following query to see the distribution of the different directions of the robot's movements, as shown in *Figure 9.4*:

```
select direction, count(*)
from chapter9_deeplearning.robot_navigation
group by 1;
```

To view the results as a bar graph, please click on the toggle **Chart** button (Chart) on the **Result** pane. Under **Traces**, click on + **Trace** (+ Trace) and set **Type** as **Bar**, **X-axis** as **Direction**, and **Y-axis** as **Count** from the dropdown. Keep **Orientation** as **Vertical**.

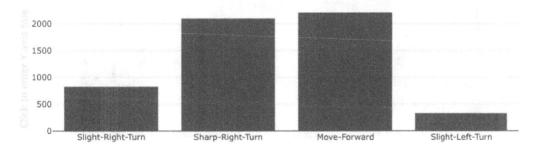

Figure 9.4 – Graph generated using Query Editor v2

You can notice that there are more *Sharp-Right-Turn* and *Move-Forward* directions than *Slight-Right-Turn* and *Slight-Left-Turn* directions. We will use these inputs to predict the future movement of the robot.

Prediction goal

Since this is a multiclass classification problem, the goal of this model is to predict which direction the robot will take next based on the 24 sensor readings.

The dataset has one ID column, which uniquely identifies a row of 24 sensor readings named `us1`, `us2`, ..., `us24`, and a `direction` variable, which has 4 values in it. The `direction` variable is the class variable that we are trying to predict.

Now let's split the dataset into a training dataset, which will be input to our model, and a test dataset, which we will use to do our predictions.

Splitting data into training and test datasets

We are going to split our table into two datasets, train and test, with an approximately 80:20 split. Let's use the `mod` function in Redshift to split our table. The `mod` function returns the remainder of two numbers. We will pass in the ID and the number 5.

To train the model, let's use `where  mod(id,5)` is not equal to 0, which represents our training set of 80%. Run the following command in Query Editor v2:

```
select direction, count(*)
from chapter9_deeplearning.robot_navigation
where mod(id,5) <> 0
group by 1;
```

In *Figure 9.5*, we see the data distribution based on ~80% of the data:

direction	count	☰
Slight-Right-Turn	662	
Sharp-Right-Turn	1680	
Move-Forward	1760	
Slight-Left-Turn	263	

Figure 9.5 – Training dataset distribution

> **Note**
> You might see a different count than we have shown. We are using Redshift's Identity function to generate the values for the id column. To be sure that the identity values are unique, Amazon Redshift skips some values when creating the identity values. Identity values are unique but the order might not match. Hence, you might see a different count but the data is 80% of the total count (5,456 rows).

The **Chart** function in Query Editor v2 depicts this in a bar chart format as shown in *Figure 9.6*:

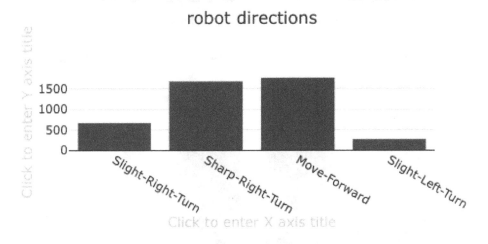

Figure 9.6 – Training set bar chart

To test the model, let's use where mod(id,5) is equal to 0, which represents our test dataset of 20%:

```
select direction, count(*) from chapter9_deeplearning.robot_navigation
where mod(id,5) = 0
group by 1;
```

In *Figure 9.7*, we see the data distribution based on ~20% of the data:

direction	count
Slight-Right-Turn	164
Sharp-Right-Turn	417
Move-Forward	445
Slight-Left-Turn	65

Figure 9.7 – Test dataset distribution

The **Chart** function in Query Editor v2 depicts this in a bar chart format as shown in *Figure 9.8*:

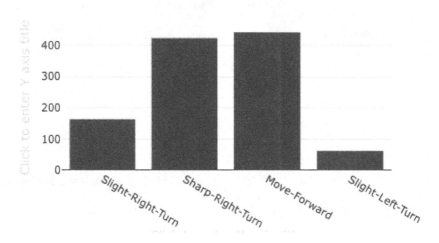

Figure 9.8 – Test data bar chart

Now that we have analyzed our data and determined how we will split it into training and test datasets, let's create our model using the MLP algorithm.

Creating a multiclass classification model using MLP

In this exercise, we are going to guide the CREATE MODEL statement to use the MLP model. You will achieve that by setting the model_type parameter to MLP. The rest of the parameters can be set to default.

Let's create a model to predict the direction of the robot:

```
CREATE MODEL chapter9_deeplearning.predict_robot_direction
from  (select
us1 ,us2 , us3 , us4 , us5 , us6 ,us7 , us8 , us9 ,
us10 ,us11 ,us12 ,us13 ,us14 ,us15 ,us16 ,us17 ,
us18 ,us19 ,us20 ,us21 , us22 ,us23 ,us24 , direction
  from chapter9_deeplearning.robot_navigation
  where mod(id,5) !=0)
target direction
function predict_robot_direction_fn
iam_role default
model_type mlp
settings (s3_bucket 'replace-with-your-s3-bucket',
max_runtime 1800);
```

The CREATE MODEL function is run with a max_runtime value of 1800 seconds. This means the maximum amount of time to train the model is 30 minutes. Training jobs often complete sooner depending on the dataset size. Since we have not set other parameters (for example, objective or problem type), Amazon SageMaker Autopilot will be doing the bulk of the work to identify the parameters for us.

Run the SHOW MODEL command to check whether model training is completed:

```
SHOW MODEL chapter9_deeplearning.predict_robot_direction;
```

Check **Model State** in *Figure 9.9*:

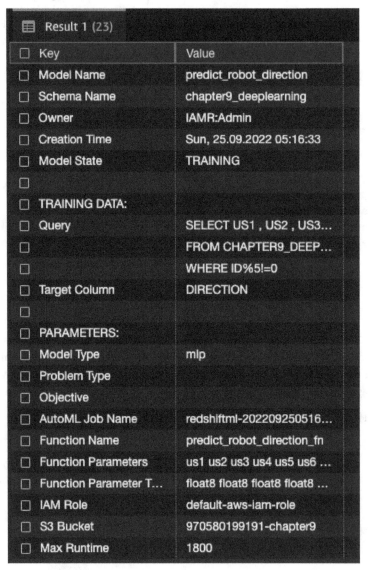

Figure 9.9 – SHOW MODEL output

From the preceding screenshot, we can see that the **Model State** field shows the status as **TRAINING**, which means the model is still under training. But notice that Redshift ML has picked up **Model Type** as **mlp**; other parameters such as **Problem Type** and **Objective** are empty now, but after the model has been trained, we will see these values.

Run the SHOW MODEL command again after some time to check whether model training is complete or not. From the following screenshot, notice that model training has finished and **Accuracy** has been selected as the objective for model evaluation. This is auto-selected by Redshift ML. Also notice that Redshift ML correctly recognized this as a multiclass classification problem:

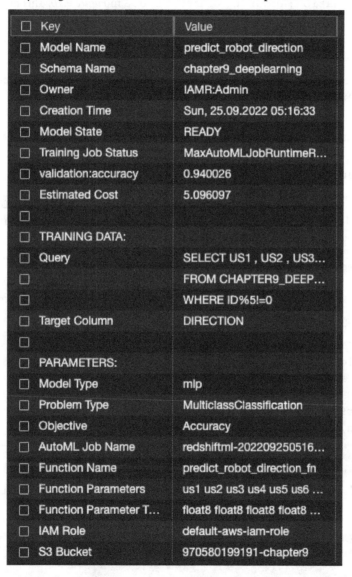

Key	Value
Model Name	predict_robot_direction
Schema Name	chapter9_deeplearning
Owner	IAMR:Admin
Creation Time	Sun, 25.09.2022 05:16:33
Model State	READY
Training Job Status	MaxAutoMLJobRuntimeR…
validation:accuracy	0.940026
Estimated Cost	5.096097
TRAINING DATA:	
Query	SELECT US1 , US2 , US3…
	FROM CHAPTER9_DEEP…
	WHERE ID%5!=0
Target Column	DIRECTION
PARAMETERS:	
Model Type	mlp
Problem Type	MulticlassClassification
Objective	Accuracy
AutoML Job Name	redshiftml-202209250516…
Function Name	predict_robot_direction_fn
Function Parameters	us1 us2 us3 us4 us5 us6 …
Function Parameter T…	float8 float8 float8 float8 …
IAM Role	default-aws-iam-role
S3 Bucket	970580199191-chapter9

Figure 9.10 – SHOW MODEL output

Now that our model has finished training, let's run predictions using the function that was created. In *Figure 9.10*, **Function Name** is `predict_robot_direction_fn` and we will refer to that in our prediction query.

Also note the `validation:accuracy` value of `.940026` in *Figure 9.10*. This means our model has an accuracy of >94%, which is very good.

> **Note**
>
> You might get a different accuracy value due to different hyperparameters selected for the algorithm in the background, and this can slightly affect accuracy.

Since our model has been successfully trained, let's run some predictions on our test dataset.

Running predictions

In this first query, we will be using the function returned by the CREATE MODEL command to compare the actual direction with our predicted directions. Run the following query in Query Editor v2 to see how many times we predicted correctly:

```
select correct, count(*)
from
(select  DIRECTION as actual, chapter9_deeplearning.predict_robot_
direction_fn (
US1,US2,US3,US4,US5,US6,US7,US8,US9,US10,US11,US12,
US13,US14,US15,US16,US17,US18,US19,US20,US21,US22,US23,US24
 ) as  predicted,
   CASE WHEN actual = predicted THEN 1::INT
          ELSE 0::INT END AS correct
from chapter9_deeplearning.robot_navigation
where MOD(id,5) =0
 ) t1
group by 1;
```

In *Figure 9.11*, we see that our model correctly predicted the robot's direction 1,033 times.

Please note that your count might be slightly different:

	correct	count
☐	1	1033
☐	0	58

Figure 9.11 – Actual directions versus predicted direction

Now, let's run a query against the test dataset to predict which direction the robot will move. Run the following query in Query Editor v2 to return the first 10 rows:

```
select  id, chapter9_deeplearning.predict_robot_direction_fn (
US1,US2,US3,US4,US5,US6,US7,US8,US9,US10,US11,US12,
US13,US14,US15,US16,US17,US18,US19,US20,US21,US22,US23,US24
 ) as  predicted_direction
from chapter9_deeplearning.robot_navigation
where MOD(id,5) <> 0
limit 10;
```

In *Figure 9.12*, we show the first 10 rows and the direction based on the ID:

id	predicted_direction
91	Slight-Right-Turn
219	Slight-Right-Turn
347	Slight-Right-Turn
603	Slight-Right-Turn
731	Slight-Right-Turn
859	Slight-Right-Turn
987	Slight-Right-Turn
1243	Sharp-Right-Turn
1371	Sharp-Right-Turn
1499	Sharp-Right-Turn

Figure 9.12 – Predicted direction by ID

Now, let's modify the query to summarize our predicted robot movements. Run the following in Query Editor v2:

```
select chapter9_deeplearning.predict_robot_direction_fn (
US1,US2,US3,US4,US5,US6,US7,US8,US9,US10,US11,US12,
US13,US14,US15,US16,US17,US18,US19,US20,US21,US22,US23,US24
 ) as  predicted_direction, count(*)
from chapter9_deeplearning.robot_navigation
where MOD(id,5) <> 0
group by 1;
```

In *Figure 9.13*, we can see that **Move-Forward** is the most popular direction, followed closely by **Sharp-Right-Turn**. Please note that your counts might differ slightly.

predicted_direction	count
Slight-Right-Turn	660
Sharp-Right-Turn	1707
Move-Forward	1737
Slight-Left-Turn	261

Figure 9.13 – Summary of predicted direction

You have now created a model using the MLP algorithm and run predictions on the test dataset.

Summary

In this chapter, we discussed deep learning models and why you need them and showed you how to create an MLP model on sensor-reading data to predict the next movement of the robot. You learned that non-linear datasets are suited for deep learning and created a multiclass classification model using the MLP algorithm.

In the next chapter, we will show you how to create a model with complete control of hyper-tuning parameters using XGBoost algorithms.

10

Creating a Custom ML Model with XGBoost

So far, all of the supervised learning models we have explored have utilized the **Amazon Redshift Auto ML** feature, which uses **Amazon SageMaker Autopilot** behind the scenes. In this chapter, we will explore how to create custom **machine learning** (**ML**) models. Training a custom model gives you the flexibility to choose the model type and the hyperparameters to use. This chapter will provide examples of this modeling technique. By the end of this chapter, you will know how to create a custom XGBoost model and how to prepare the data to train your model using Redshift SQL.

In this chapter, we will go through the following main topics:

- Introducing XGBoost
- Introducing an XGBoost use case
- XGBoost model with Auto off feature

Technical requirements

This chapter requires a web browser and access to the following:

- An AWS account
- An Amazon Redshift Serverless endpoint
- Amazon Redshift Query Editor v2

You can find the code used in this chapter here:

```
https://github.com/PacktPublishing/Serverless-Machine-Learning-with-
Amazon-Redshift/blob/main/CodeFiles/chapter10/chapter10.sql
```

Introducing XGBoost

XGBoost gets its name because it is built on the **Gradient Boosting** framework. Using a tree-boosting technique provides a fast method for solving ML problems. As you have seen in previous chapters, you can specify the model type, which can help speed up model training since **SageMaker Autopilot** does not have to determine which model type to use.

You can learn more about XGBoost here: `https://docs.aws.amazon.com/sagemaker/latest/dg/xgboost.html`.

When you create a model with Redshift ML and specify XGBoost as the model type, and optionally specify AUTO OFF, this turns off SageMaker Autopilot and you have more control of model tuning. For example, you can specify the hyperparameters you wish to use. You will see an example of this in the *Creating a binary classification model using XGBoost* section.

You will have to perform preprocessing when you set **AUTO** to **OFF**. Carrying out the preprocessing ensures we will get the best possible model and is also necessary since all inputs must be numeric when you set **AUTO** to **OFF**, for example, by making sure data is cleansed, categorical variables are encoded, and numeric variables are standardized. You will also need to identify the type of problem that you have and select an appropriate model to train. You will be able to create train and test datasets and evaluate models yourself. You also have the ability to tune the hyperparameters. In summary, you get total control of the end-to-end ML model training and building.

By using XGBoost with Amazon Redshift ML, you can solve both regression and classification problems. You also can specify the learning objective of your model. For example, if you are solving a binary classification problem, you would choose `binary:logistic` as your objective or use `multi:softmax` for multi-class classification problems.

At the time of writing this book, the supported learning objectives are `reg:squarederror`, `reg:squaredlogerror`, `reg:logistic`, `reg:pseudohubererror`, `reg:tweedie`, `binary:logistic`, `binary:hinge`, and `multi:softmax`.

For more information about these objectives, see the *Learning Task Parameters* section of the XGBoost documentation here: `https://xgboost.readthedocs.io/en/latest/parameter.html#learning-task-parameters`.

Now that you have learned what XGBoost is, we will take a look at a use case where we can apply XGBoost and solve a common business problem using binary classification.

Introducing an XGBoost use case

In this section, we will be discussing a use case where we want to predict whether credit card transactions are fraudulent. We will be going through the following steps:

- Defining the business problem
- Uploading, analyzing, and preparing data for training
- Splitting data into training and testing datasets
- Preprocessing the input variables

Defining the business problem

In this section, we will use a credit card payment transaction dataset to build a binary classification model using XGBoost in Redshift ML. This dataset contains customer and terminal information along with the date and amount related to the transaction. This dataset also has some derived fields based on **recency**, **frequency**, and **monetary** numeric features, along with a few categorical variables, such as whether a transaction occurred during the weekend or at night. Our goal is to identify whether a transaction is fraudulent or non-fraudulent. This use case is taken from `https://github.com/Fraud-Detection-Handbook/fraud-detection-handbook`. Please refer to the GitHub repository to learn more about this data generation process.

> **Dataset citation**
>
> *Reproducible Machine Learning for Credit Card Fraud Detection - Practical Handbook*, Le Borgne, Yann-Aël and Siblini, Wissam and Lebichot, Bertrand and Bontempi, Gianluca, `https://github.com/Fraud-Detection-Handbook/fraud-detection-handbook`, 2022, Université Libre de Bruxelles

Now, we will load our dataset into Amazon Redshift ML and prepare it for model training.

Uploading, analyzing, and preparing data for training

Before we begin, let's first connect to Redshift as an admin or database developer and then load data into Amazon Redshift.

In the following steps, you will create a schema for all of the tables and objects needed for this exercise, which involves creating all the needed tables, loading data, and creating the views used for data transformations.

Navigate to Query Editor v2, connect to the serverless endpoint, and then connect to the **dev** database, as shown in the following screenshot:

Figure 10.1 – Connect to Query Editor v2

1. Execute the following step to create the schema. This schema will be used for all objects and models created in this chapter:

    ```
    CREATE SCHEMA  chapter10_xgboost;
    ```

2. Next, copy the following SQL statement into Query Editor v2 to create the table for hosting the customer payment transaction history, which we will load in the subsequent step:

    ```
    create table chapter10_xgboost.cust_payment_tx_history
    (
    transaction_id integer,
    tx_datetime timestamp,
    customer_id integer,
    terminal_id integer,
    tx_amount decimal(9,2),
    tx_time_seconds integer,
    tx_time_days integer,
    tx_fraud integer,
    tx_fraud_scenario integer,
    tx_during_weekend integer,
    tx_during_night integer,
    customer_id_nb_tx_1day_window decimal(9,2),
    customer_id_avg_amount_1day_window decimal(9,2),
    customer_id_nb_tx_7day_window decimal(9,2),
    customer_id_avg_amount_7day_window decimal(9,2),
    customer_id_nb_tx_30day_window decimal(9,2),
    customer_id_avg_amount_30day_window decimal(9,2),
    terminal_id_nb_tx_1day_window decimal(9,2),
    terminal_id_risk_1day_window decimal(9,2),
    terminal_id_nb_tx_7day_window decimal(9,2),
    terminal_id_risk_7day_window decimal(9,2),
    ```

```
terminal_id_nb_tx_30day_window decimal(9,2),
terminal_id_risk_30day_window decimal(9,2)
)
;
```

3. Now that you have created the table, you can execute the following command in Query Editor v2 to load the table:

```
copy chapter10_xgboost.cust_payment_tx_history
from 's3://packt-serverless-ml-redshift/chapter10/credit_card_
transactions_transformed_balanced.csv'
iam_role default
ignoreheader 1
csv region 'eu-west-1';
```

4. Now that you have loaded the data, it's a good practice to sample some data to make sure our data is loaded properly. Run the following query to sample 10 records:

```
select * from
chapter10_xgboost.cust_payment_tx_history
limit 10;
```

In the following screenshot, we can see that we have loaded the data correctly with a sampling of different transaction IDs:

transaction_id	tx_datetime	customer_id	terminal_id	tx_amount
48262	2022-06-06 02:35:29	2497	9180	43.6
48263	2022-06-06 02:35:40	1534	4771	2.14
48264	2022-06-06 02:36:01	1593	4770	46.14
48265	2022-06-06 02:36:05	4069	7408	155.28
48266	2022-06-06 02:37:01	655	6930	174.5
48267	2022-06-06 02:37:11	684	6893	32.33
48268	2022-06-06 02:37:19	3134	2287	15.52

Figure 10.2 – Data sample

As discussed in earlier chapters, the target variable is the value that we are trying to predict in our model. In our use case, we are trying to predict whether a transaction is fraudulent. In our dataset, this is the tx_fraud attribute, which is our target. Let us check our table to see how many transactions were flagged as fraudulent.

Run the following command in Query Editor v2:

```
select tx_fraud, count(*)
from chapter10_xgboost.cust_payment_tx_history
group by 1;
```

We identify fraudulent transactions in our dataset as those with a `tx_fraud` value of 1. We have identified 14,681 transactions as fraudulent in our dataset. Conversely, a `tx_fraud` value of 0 indicates that a transaction is not fraudulent:

tx_fraud	count
0	25487
1	14681

Figure 10.3 – Fraudulent transactions

Let us look at the trend of fraudulent and non-fraudulent transactions over the months. We want to analyze whether there are any unusual spikes in fraudulent transactions.

Run the following SQL command in Query Editor v2:

```
select to_char(tx_datetime, 'yyyymm') as yearmonth,
sum(case when tx_fraud = 1 then 1 else 0 end) fraud_tx,
sum(case when tx_fraud = 0 then 1 else 0 end) non_fraud_tx,
count(*) as total_tx,
(fraud_tx::decimal(10,2 ) / total_tx::decimal(10,2) ) *100 as fraud_
txn_pct
from chapter10_xgboost.cust_payment_tx_history
group by yearmonth
order by yearmonth
```

Notice that fraudulent transactions increased by nearly 8 percent in 202207 over 202206:

yearmonth	fraud_tx	non_fraud_tx	total_tx	fraud_txn_pct
202206	1702	4225	5927	28.716045216
202207	2639	4666	7305	36.125941136
202208	2504	4350	6854	36.533411146
202209	2620	3821	6441	40.676913522
202210	2669	4100	6769	39.429753287
202211	2547	4325	6872	37.063445867

Figure 10.4 – Fraudulent transaction trends

Now that we have loaded our data, let's get our data prepared for model training by splitting the data into train and test datasets. The training data is used to train the model and the testing data is used to run our prediction queries.

Splitting data into train and test datasets

To train the model, we will have transactions that are older than 2022-10-01, which is ~ 80 percent of the transactions.

To test the model, we will use transactions from after 2022-09-30, which is 20 percent of the transactions.

Preprocessing the input variables

We have a combination of numeric and categorical variables in our input fields. We need to preprocess the categorical variables into one-hot-encoded values and standardize the numeric variables. Since we will be using **AUTO OFF**, SageMaker does not automatically preprocess the data. Hence, it is important to transform various numeric, datetime, and categorical features.

Categorical features (also referred to as nominal) have distinct categories or levels. These can be categories without an order to them, such as country or gender. Or they can have an order such as level of education (also referred to as ordinal).

Since ML models need to operate on **numeric variables**, we need to apply ordinal encoding or one-hot encoding.

To make things easier, we have created the following view to take care of the transformation logic. This view is somewhat lengthy, but actually, what the view is doing is quite simple:

- Calculating the transaction time in seconds and days

- Applying one-hot encoding by assigning 0 or 1 to classify transactions as weekday, weekend, daytime, or nighttime (such as TX_DURING_WEEKEND or TX_DURING_NIGHT)

- Applying window functions to transactions so that we make it easy to visualize the data in 1-day, 7-day, and 30-day intervals

Execute the following SQL command in Query Editor v2 to create the view by applying the transformation logic:

```
create view chapter10_xgboost.credit_payment_tx_history_scaled
as
select
transaction_id, tx_datetime, customer_id, terminal_id,
tx_amount ,
( (tx_amount - avg(tx_amount) over()) /  cast(stddev_pop(tx_amount)
```

```
over() as dec(14,2)) ) s_tx_amount,
tx_time_seconds ,
  ( (tx_time_seconds - avg(tx_time_seconds) over()) / cast(stddev_
pop(tx_time_seconds) over() as dec(14,2)) ) s_tx_time_seconds,
tx_time_days  ,
  ( (tx_time_days - avg(tx_time_days) over()) / cast(stddev_pop(tx_
time_days) over() as dec(14,2)) ) s_tx_time_days,
tx_fraud  ,
  tx_during_weekend ,
case when tx_during_weekend = 1 then 1 else 0 end as tx_during_
weekend_ind,
case when tx_during_weekend = 0 then 1 else 0 end tx_during_weekday_
ind,
tx_during_night,
case when tx_during_night = 1 then 1 else 0 end as tx_during_night_
ind,
case when tx_during_night = 0 then 1 else 0 end as tx_during_day_ind,
customer_id_nb_tx_1day_window ,
  ( (customer_id_nb_tx_1day_window - avg(customer_id_nb_tx_1day_
window) over()) / cast(stddev_pop(customer_id_nb_tx_1day_window)
over() as dec(14,2)) ) s_customer_id_nb_tx_1day_window,
customer_id_avg_amount_1day_window  ,
  ( (customer_id_avg_amount_1day_window - avg(customer_id_avg_
amount_1day_window) over()) / cast(stddev_pop(customer_id_avg_
amount_1day_window) over() as dec(14,2)) ) s_customer_id_avg_
amount_1day_window,
customer_id_nb_tx_7day_window ,
  ( (customer_id_nb_tx_7day_window - avg(customer_id_nb_tx_7day_
window) over()) / cast(stddev_pop(customer_id_nb_tx_7day_window)
over() as dec(14,2)) ) s_customer_id_nb_tx_7day_window,
customer_id_avg_amount_7day_window  ,
  ( (customer_id_avg_amount_7day_window - avg(customer_id_avg_
amount_7day_window) over()) / cast(stddev_pop(customer_id_avg_
amount_7day_window) over() as dec(14,2)) ) s_customer_id_avg_
amount_7day_window,
customer_id_nb_tx_30day_window  ,
  ( (customer_id_nb_tx_30day_window - avg(customer_id_nb_tx_30day_
window) over()) / cast(stddev_pop(customer_id_nb_tx_30day_window)
over() as dec(14,2)) ) s_customer_id_nb_tx_30day_window,
customer_id_avg_amount_30day_window ,
  ( (customer_id_avg_amount_30day_window - avg(customer_id_avg_
amount_30day_window) over()) / cast(stddev_pop(customer_id_avg_
amount_30day_window) over() as dec(14,2)) ) s_customer_id_avg_
amount_30day_window,
terminal_id_nb_tx_1day_window ,
  ( (terminal_id_nb_tx_1day_window - avg(terminal_id_nb_tx_1day_
window) over()) / cast(stddev_pop(terminal_id_nb_tx_1day_window)
over() as dec(14,2)) ) s_terminal_id_nb_tx_1day_window,
```

```
terminal_id_risk_1day_window  ,
   ( (terminal_id_risk_1day_window - avg(terminal_id_risk_1day_window)
over()) / cast(stddev_pop(terminal_id_risk_1day_window) over() as
dec(14,2)) ) s_terminal_id_risk_1day_window,
terminal_id_nb_tx_7day_window ,
   ( (terminal_id_nb_tx_7day_window - avg(terminal_id_nb_tx_7day_
window) over()) / cast(stddev_pop(terminal_id_nb_tx_7day_window)
over() as dec(14,2)) ) s_terminal_id_nb_tx_7day_window,
terminal_id_risk_7day_window  ,
   ( (terminal_id_risk_7day_window - avg(terminal_id_risk_7day_window)
over()) / cast(stddev_pop(terminal_id_risk_7day_window) over() as
dec(14,2)) ) s_terminal_id_risk_7day_window,
terminal_id_nb_tx_30day_window  ,
   ( (terminal_id_nb_tx_30day_window - avg(terminal_id_nb_tx_30day_
window) over()) / cast(stddev_pop(terminal_id_nb_tx_30day_window)
over() as dec(14,2)) ) s_terminal_id_nb_tx_30day_window,
terminal_id_risk_30day_window ,
   ( (terminal_id_risk_30day_window - avg(terminal_id_risk_30day_
window) over()) / cast(stddev_pop(terminal_id_risk_30day_window)
over() as dec(14,2)) ) s_terminal_id_risk_30day_window
from
chapter10_xgboost.cust_payment_tx_history;
```

Now that the view is created, let's sample 10 records.

Execute the following command in Query Editor v2:

```
SELECT * from chapter10_XGBoost.credit_payment_tx_history_scaled limit
10;
```

We can see some of our transformed values, such as `tx_time_seconds` and `txn_time_days`, in the following screenshot:

tx_time_seconds	s_tx_time_seconds	tx_time_days	s_tx_time_days	tx_fraud	tx
441329	-1.4993089198786	5	-1.4922553834529	0	0
441340	-1.4993065181455	5	-1.4922553834529	0	0
441361	-1.4993019330188	5	-1.4922553834529	0	0
441365	-1.4993010596614	5	-1.4922553834529	0	0
441421	-1.4992888326568	5	-1.4922553834529	0	0
441431	-1.4992866492632	5	-1.4922553834529	0	0
441439	-1.4992849025482	5	-1.4922553834529	0	0
441484	-1.4992750772767	5	-1.4922553834529	0	0

Figure 10.5 – Transformed data

Now, let's quickly review why we needed to create this view:

- Since we are using XGBoost with Auto OFF, we must do our own data preprocessing and feature engineering

- We applied one-hot encoding to our categorical variables

- We scaled our numeric variables

Here is a summary of the view logic:

- The target variable we used is `TX_FRAUD`

- The categorical variables we used are `TX_DURING_WEEKEND_IND`, `TX_DURING_WEEKDAY_IND`, `TX_DURING_NIGHT_IND`, and `TX_DURING_DAY_IND`

- The scaled numeric variables are `s_customer_id_nb_tx_1day_window`, `s_customer_id_avg_amount_1day_window`, `s_customer_id_nb_tx_7day_window`, `s_customer_id_avg_amount_7day_window`, `s_customer_id_nb_tx_30day_window`, `s_customer_id_avg_amount_30day_window`, `s_terminal_id_nb_tx_1day_window`, `s_terminal_id_risk_1day_window`, `s_terminal_id_nb_tx_7day_window`, `s_terminal_id_risk_7day_window`, `s_terminal_id_nb_tx_30day_window`, and `s_terminal_id_risk_30day_window`

You have now completed data preparation and are ready to create your model!

Creating a model using XGBoost with Auto Off

In this exercise, we are going to create a custom binary classification model using the XGBoost algorithm. You can achieve this by setting **AUTO off**. Here are the parameters that are available:

- **AUTO OFF**

- **MODEL_TYPE**

- **OBJECTIVE**

- **HYPERPARAMETERS**

For the complete list of hyperparameter values that are available and their defaults, please read the documentation found here:

`https://docs.aws.amazon.com/redshift/latest/dg/r_create_model_use_cases.html#r_auto_off_create_model`

Now that you have a basic understanding of the parameters available with XGBoost, you can create the model.

Creating a binary classification model using XGBoost

Let's create a model to predict whether a transaction is fraudulent or non-fraudulent. As you learned in the previous chapters, creating models with Amazon Redshift ML is simply done by running a SQL command that creates a function. As inputs (or features), you will be using the attributes from the view that you created in the previous section. You will specify `tx_fraud` as the target and give the function name, which you will use later in your prediction queries. Additionally, you will specify hyperparameters to do your own model tuning. Let's begin!

Execute the following commands in Query Editor v2. The following is a code snippet; you may retrieve the full code from the following URL:

`https://github.com/PacktPublishing/Serverless-Machine-Learning-with-Amazon-Redshift/blob/main/chapter10.sql`

```
drop model chapter10_XGBoost.cust_cc_txn_fd_xg;

 create model chapter10_xgboost.cust_cc_txn_fd_xg
from (
select
  s_tx_amount,
tx_fraud,
...
  from chapter10_xgboost.payment_tx_history_scaled
  where cast(tx_datetime as date) between '2022-06-01' and '2022-09-30'
)
target tx_fraud
function fn_customer_cc_fd_xg
iam_role default
auto off
model_type xgboost
objective 'binary:logistic'
preprocessors 'none'
hyperparameters default except (num_round '100')
settings (
  s3_bucket '<<your-s3-bucket>>',

            s3_garbage_collect off,
            max_runtime 1500
                  );
```

The CREATE MODEL function is going to invoke the XGBoost algorithm and train a binary classification model. We have set **AUTO off**, which means Autopilot is not going to perform any tasks for us. We are customizing the model to be a binary classifier using preprocessed data. We also set the num_round hyperparameter value to 100, which is the number of rounds to run the training.

Now, let's run SHOW MODEL to see whether model training is completed. Run the following command in Query Editor v2:

```
SHOW MODEL  chapter10_XGBoost.cust_cc_txn_fd_xg;
```

Note **Model State** in the following screenshot, which shows your model is still training:

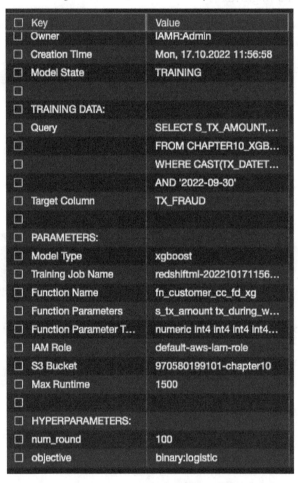

Key	Value
Owner	IAMR:Admin
Creation Time	Mon, 17.10.2022 11:56:58
Model State	TRAINING
TRAINING DATA:	
Query	SELECT S_TX_AMOUNT,...
	FROM CHAPTER10_XGB...
	WHERE CAST(TX_DATET...
	AND '2022-09-30'
Target Column	TX_FRAUD
PARAMETERS:	
Model Type	xgboost
Training Job Name	redshiftml-202210171156...
Function Name	fn_customer_cc_fd_xg
Function Parameters	s_tx_amount tx_during_w...
Function Parameter T...	numeric int4 int4 int4...
IAM Role	default-aws-iam-role
S3 Bucket	970580199101-chapter10
Max Runtime	1500
HYPERPARAMETERS:	
num_round	100
objective	binary:logistic

Figure 10.6 – Show model output

From the preceding screenshot, we notice that the value of **Model State** is TRAINING, which is self-explanatory – the model is still training. You will also see that Redshift ML has picked up the parameters we supplied in the CREATE MODEL statement – **Model Type** is set to **xgboost**. **objective** is set to **binary:logistic** and the **num_round** parameter is set to **100**.

When you have a custom model with **AUTO OFF** and specify the hyperparameters, the model can be trained much faster. This model will usually finish in under 10 minutes.

Run the SHOW MODEL command again after 10 minutes to check whether model training is complete or not. As you can see from the following screenshot, model training has completed and the **train:error** field reports the error rate. Most datasets have a threshold of .5, so our value of **0.051870** is very good, as seen in the following screenshot:

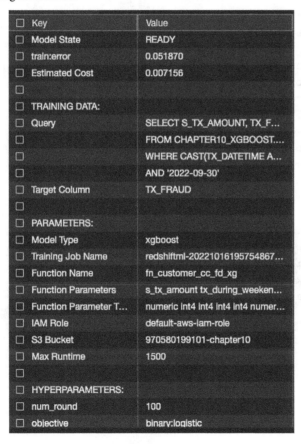

Key	Value
Model State	READY
train:error	0.051870
Estimated Cost	0.007156
TRAINING DATA:	
Query	SELECT S_TX_AMOUNT, TX_F...
	FROM CHAPTER10_XGBOOST....
	WHERE CAST(TX_DATETIME A...
	AND '2022-09-30'
Target Column	TX_FRAUD
PARAMETERS:	
Model Type	xgboost
Training Job Name	redshiftml-20221016195754867...
Function Name	fn_customer_cc_fd_xg
Function Parameters	s_tx_amount tx_during_weeken...
Function Parameter T...	numeric int4 int4 int4 numer...
IAM Role	default-aws-iam-role
S3 Bucket	970580199101-chapter10
Max Runtime	1500
HYPERPARAMETERS:	
num_round	100
objective	binary:logistic

Figure 10.7 – SHOW MODEL output

Now, your model is complete and has a good score based on `score - train_error`, which is `0.051870`. You are now ready to use it for predictions.

Generating predictions and evaluating model performance

Run the following query in Query Editor v2, which will compare the actual `tx_fraud` value with the `predicted_tx_fraud` value:

```
select
tx_fraud ,
fn_customer_cc_fd_xg(
s_tx_amount,
tx_during_weekend_ind,
tx_during_weekday_ind,
tx_during_night_ind,
tx_during_day_ind,
s_customer_id_nb_tx_1day_window,
s_customer_id_avg_amount_1day_window,
s_customer_id_nb_tx_7day_window,
s_customer_id_avg_amount_7day_window,
s_customer_id_nb_tx_30day_window,
s_customer_id_avg_amount_30day_window,
s_terminal_id_nb_tx_1day_window,
s_terminal_id_risk_1day_window,
s_terminal_id_nb_tx_7day_window,
s_terminal_id_risk_7day_window,
s_terminal_id_nb_tx_30day_window,
s_terminal_id_risk_30day_window)
from chapter10_xgboost.credit_payment_tx_history_scaled
where cast(tx_datetime as date) >= '2022-10-01'
;
```

The following screenshot shows the sample output. In this screenshot, our predicted values are the same as the actual values:

tx_fraud	fn_customer_cc_fd_xg
0	0
0	0
0	0
0	0
0	0
0	0
0	0
0	0
0	0
0	0
0	0
0	0
0	0

Figure 10.8 – Inference query output

Since we did not get the F1 value for our model from Redshift ML, let's calculate it. We will create a view that contains the logic to accomplish this:

```
--drop view if exists chapter10_xgboost.fraud_tx_conf_matrix;
create or replace view chapter10_xgboost.fraud_tx_conf_matrix
as
select
transaction_id,tx_datetime,customer_id,tx_amount,terminal_id, tx_
fraud,
  fn_customer_cc_fd_xg(
  s_tx_amount,
tx_during_weekend_ind,
tx_during_weekday_ind,
tx_during_night_ind,
tx_during_day_ind,
s_customer_id_nb_tx_1day_window,
s_customer_id_avg_amount_1day_window,
s_customer_id_nb_tx_7day_window,
```

```
s_customer_id_avg_amount_7day_window,
s_customer_id_nb_tx_30day_window,
s_customer_id_avg_amount_30day_window,
s_terminal_id_nb_tx_1day_window,
s_terminal_id_risk_1day_window,
s_terminal_id_nb_tx_7day_window,
s_terminal_id_risk_7day_window,
s_terminal_id_nb_tx_30day_window,
s_terminal_id_risk_30day_window)
as prediction,
case when tx_fraud  =1 and prediction = 1 then 1 else 0 end
truepositives,
case when tx_fraud =0 and prediction = 0 then 1 else 0 end
truenegatives,
case when tx_fraud =0 and prediction = 1 then 1 else 0 end
falsepositives,
case when tx_fraud =1 and prediction = 0 then 1 else 0 end
falsenegatives
    from chapter10_xgboost.credit_payment_tx_history_scaled
    where cast(tx_datetime as date) >= '2022-10-01';
```

Run the following SQL command in Query Editor v2 to check the F1 score that we calculated in the view:

```
select
sum(truepositives+truenegatives)*1.00/(count(*)*1.00) as accuracy,--
accuracy of the model,
sum(falsepositives+falsenegatives)*1.00/count(*)*1.00 as error_rate,
--how often model is wrong,
sum(truepositives)*1.00/sum (truepositives+falsenegatives) *1.00 as
tpr, --or recall how often corrects are rights,
sum(falsepositives)*1.00/sum (falsepositives+truenegatives )*1.00 fpr,
--or fall-out how often model said yes when it is no,
sum(truenegatives)*1.00/sum (falsepositives+truenegatives)*1.00 tnr,
--or specificity, how often model said no when it is yes,
sum(truepositives)*1.00 / (sum (truepositives+falsepositives)*1.00) as
precision, -- when said yes how it is correct,
2*((tpr*precision)/ (tpr+precision) ) as f_score --weighted avg of tpr
& fpr
from chapter10_xgboost.fraud_tx_conf_matrix
;
```

You can see our accuracy is 90 percent and our F1 score is 87 percent, which are both very good. Additionally, our confusion matrix values tell us how many times we correctly predicted `True` and correctly predicted `False`:

accuracy	error_rate	tpr	fpr	trv	precision	f_score
0.907704713730664	0.0922952862693350	0.79275306748466257	0.02112759843916913	0.97887240356083086	0.958729422675631	0.86787

Figure 10.9 – F1 score

Now, let's check actual versus prediction counts. Run the following query in Query Editor v2:

```
select tx_fraud,prediction, count(*)
from chapter10_xgboost.fraud_tx_conf_matrix
group by tx_fraud,prediction;
```

The output in the following screenshot shows, for a given value, what our prediction was compared to the actual value and the count of those records. Our model incorrectly predicted a fraudulent transaction 178 times and incorrectly predicted a non-fraudulent transaction 1,081 times:

Result 1 (4)

tx_fraud	prediction	count
0	0	8247
0	1	178
1	1	4135
1	0	1081

Figure 10.10 – Confusion matrix

This demonstrates how Redshift ML can help you confidently predict whether a transaction is fraudulent.

Summary

In this chapter, you learned what XGBoost is and how to apply it to a business problem. You learned how to specify your own hyperparameters when using the **Auto Off** option and how to specify the objective for a binary classification problem. Additionally, you learned how to do your own data preprocessing and calculate the F1 score to validate the model performance.

In the next chapter, you will learn how to bring your own models from Amazon SageMaker for in-database or remote inference.

11

Bringing Your Own Models for Database Inference

In this book, we've covered the process of training models natively using **Redshift Machine Learning (Redshift ML)**. However, there may be instances where you need to utilize models built outside of Redshift. To address this, Redshift ML offers the **Bring Your Own Model (BYOM)** feature, allowing users to integrate their Amazon SageMaker machine learning models with Amazon Redshift. This feature facilitates making predictions and performing other machine learning tasks on data stored in the warehouse, without requiring data movement.

BYOM offers two approaches: **local inference** and **remote inference**. In this chapter, we'll delve into the workings of BYOM and explore the various options available for creating and integrating BYOM. You'll be guided through the process of building a machine learning model in Amazon SageMaker, and subsequently, employing Redshift ML's BYOM feature to bring that model to Redshift. Moreover, you'll learn how to apply these models to the data stored in Redshift's data warehouse to make predictions.

By the end of this chapter, you'll be proficient in bringing Amazon SageMaker-created models and executing predictions within Amazon Redshift. Utilizing BYOM, you can deploy models such as **XGBoost** and a **multilayer perceptron (MLP)** to Redshift ML. Once a pre-trained model is deployed on Redshift ML, you can run inferences locally on Redshift without relying on a SageMaker endpoint or SageMaker Studio. This simplicity empowers data analysts to conduct inference on new data using models created externally to Redshift, eliminating concerns about accessing SageMaker's services.

This method significantly speeds up the delivery of machine learning models created outside of Redshift to the data team. Furthermore, since Redshift ML interacts with native Redshift SQL, the user experience for the data team remains consistent with other data analysis work performed on the data warehouse.

In this chapter, we will go through the following main topics:

- Benefits of BYOM

- Supported model types

- BYOM for local inference

- BYOM for remote inference

Technical requirements

This chapter requires a web browser and access to the following:

- An AWS account

- An Amazon Redshift Serverless endpoint

- An Amazon SageMaker notebook

- Amazon Redshift Query Editor v2

- Completing the *Getting started with Amazon Redshift Serverless* section in *Chapter 1*

You can find the code used in this chapter here:

`https://github.com/PacktPublishing/Serverless-Machine-Learning-with-Amazon-Redshift`

The data files required for this chapter are located in a public S3 bucket: `s3://packt-serverless-ml-redshift/`.

Let's begin!

Benefits of BYOM

With Amazon Redshift ML, you can use an existing ML model built in Amazon SageMaker and use it in Redshift without having to retrain it. To use BYOM, you need to provide model artifacts or a SageMaker endpoint, which takes a batch of data and returns predictions. BYOM is useful in cases where a machine learning model is not yet available in Redshift ML, for example, at the time of writing this book, a Random Cut Forest model is not yet available in Redshift ML, so you can build this model in SageMaker and easily bring it to Redshift and then use it against the data stored in Redshift.

Here are some specific benefits of using Redshift ML with your own ML model:

- **Improved efficiency**: By using an existing ML model, you can save time and resources that would otherwise be spent on training a new model

- **Easy integration**: Redshift ML makes it easy to integrate your ML model into your data pipeline, allowing you to use it for real-time predictions or batch predictions

- **Scalability**: Redshift ML is built on top of the highly scalable and performant Amazon Redshift data warehouse, so you can use your ML model to make predictions on large datasets without worrying about performance issues

Supported model types

Amazon Redshift ML supports a wide range of machine learning models through the BYOM feature. Some common types of models that can be used with BYOM include the following:

- **Linear regression models**: These models are like number predictors. They take into account several factors or features and use them to guess a specific numerical outcome. For example, if you want to predict the price of a house, a linear regression model would consider factors such as the size of the house, the number of rooms, and the location to estimate the house's price.

- **Logistic regression models**: These models are binary outcome predictors. Instead of guessing numbers, they answer *yes* or *no* questions or make *0/1* predictions. For instance, if you want to predict whether a student will pass or fail an exam, a logistic regression model would consider factors such as the student's study hours, previous test scores, and attendance to determine the likelihood of passing the exam.

- **Decision tree models**: These are used to make predictions based on a tree-like structure. Think of it like a decision-making tree for predictions. You start at the top and follow branches based on known features. At each branch, you make a decision based on a feature and keep going until you reach a final prediction at the leaves. It's a step-by-step process to find the most likely outcome.

- **Random forest models**: These are ensembles of decision trees. Groups of decision trees work together. Each tree is trained on a different part of the data. To make a prediction, all the trees give their answers, and their predictions are averaged to get the final result. It's like taking the opinions of multiple trees to make a more accurate guess.

- **Gradient boosting models**: These are also ensembles of decision trees, These are groups of decision trees that work together, but here, unlike in a random forest model, the trees are trained one after the other, and each tree tries to fix the mistakes of the previous one. They learn from each other's errors and become better as a team. It's like a learning process where they keep improving until they make good predictions together.

- **Neural network models**: These are complex, multi-layered models that are able to learn complex patterns in data. These models are capable of learning intricate patterns in data. They operate using a process of information analysis, discovering underlying correlations similar to the functioning of interconnected neurons in the human brain. Through extensive training and exposure to diverse datasets, the model refines its ability to decipher complex patterns, making it proficient in uncovering intricate relationships within new data.

- **Support vector machines (SVMs)**: SVMs are powerful classifiers, acting like incredibly intelligent dividers. Imagine a 3D space with points representing different things. SVMs determine the most optimal way to draw a line or plane, called a hyperplane, that perfectly separates two distinct groups of points. It's as if they possess an extraordinary ability to find the perfect boundary, ensuring the two groups are kept as far apart as possible, such as drawing an invisible but flawless line that keeps everything perfectly organized on each side.

These are just a few examples of the types of models that can be used with BYOM in Amazon Redshift. In general, any model that can be represented as a set of model artifacts and a prediction function can be used with BYOM in Redshift.

We have learned what Redshift ML BYOM is and its benefits. In the next section, you will create a BYOM local inference model.

Creating the BYOM local inference model

With BYOM local inference, the machine learning model and its dependencies are packaged into a group of files and deployed to Amazon Redshift where the data is stored, allowing users to make predictions on the stored data. Model artifacts and their dependencies are created when a model is trained and created on the Amazon SageMaker platform. By deploying the model directly onto the Redshift service, you are not moving the data over the network to another service. Local inference can be useful for scenarios where the data is sensitive or requires low latency predictions.

Let's start working on creating the BYOM local inference model.

Creating a local inference model

To create the BYOM local inference model, the first step involves training and validating an Amazon SageMaker model. For this purpose, we will train and validate an XGBoost linear regression machine learning model on Amazon SageMaker. Follow the instructions found here to create the Amazon SageMaker model:

```
https://github.com/aws/amazon-sagemaker-examples/blob/main/
introduction_to_amazon_algorithms/xgboost_abalone/xgboost_abalone.
ipynb
```

After you have followed the instructions given at the preceding URL, validate the model by running prediction functions. Now, let's move on to the next steps. After successfully generating the predictions, we will create the Redshift ML model. Using the same notebook, let's run a few commands to set some parameters.

Creating the model and running predictions on Redshift

Now, validate the model by running prediction functions.

With the model trained and validated in SageMaker, it's time to import it into Redshift. In the next section, using the same SageMaker notebook, we will set up the required parameters to build the Redshift CREATE MODEL statement. You will use this statement in Query Editor v2 to create your model in Redshift ML, enabling you to perform local inference on the data stored in the Redshift cluster with the integrated SageMaker model.

Setting up the parameters

Before setting up the parameters, run the following command in Query Editor v2 to create the schema for this chapter:

```
Create schema chapter11_byom;
```

The first step of this process is setting up the following parameter values:

- S3_BUCKET is used to store Redshift ML artifacts.
- MODEL_PATH is the S3 location of the model artifact of the Amazon SageMaker model. Optionally, you can print model_data using the print function in Python and look at the artifact location.
- REDSHIFT_IAM_ROLE is the cluster role:

```
#provide your s3 bucket here
S3_BUCKET='Redshift ML s3 bucket name'

#provide the model path, this is coming from the model_data parameter
MODEL_PATH=model_data
#Provide Redshift cluster attached role ARN
REDSHIFT_IAM_ROLE = 'Redshift Cluster IAM Role'
```

Next, we will generate the CREATE MODEL statement that you are going to run on Redshift.

Generating the CREATE MODEL statement

Execute the code provided here in a Jupyter notebook to automatically generate the CREATE MODEL statement:

```
sql_text=("drop model if exists predict_abalone_age; \
  CREATE MODEL chapter11_byom.predict_abalone_age \
FROM '{}' \
FUNCTION predict_abalone_age ( int, int, float,
float,float,float,float,float,float) \
RETURNS int \
IAM_ROLE '{}' \
settings( S3_BUCKET '{}') \
")
print (sql_text.format(model_data,REDSHIFT_IAM_ROLE, S3_BUCKET))
```

The output of the preceding statement is the CREATE MODEL statement that you are going to run in Query Editor v2. Please copy the statement and head over to Query Editor v2 to perform the remaining steps.

Running local inference on Redshift

The following is the CREATE MODEL statement. You should have a similar one generated, where FROM, IAM_ROLE, and S3_BUCKET have different values:

```
CREATE MODEL chapter11_byom.predict_abalone_age
FROM 's3://redshift-ml-22-redshiftmlbucket-1cckvqgktpfe0/sagemaker/
DEMO-xgboost-abalone-default/single-xgboost/DEMO-xgboost-
regression-2022-12-31-01-45-30/output/model.tar.gz'
FUNCTION predict_abalone_age ( int, int, float,
float,float,float,float,float,float) RETURNS int IAM_ROLE
'arn:aws:iam::215830312345:role/spectrumrs'
settings( S3_BUCKET 'redshift-ml-22-redshiftmlbucket-1cckvqgktpfe0') ;
```

In the preceding command, the FROM clause takes model_data as input, which contains the SageMaker model artifacts. When this command is run, Amazon Redshift ML compiles the model, deploys it to Redshift, and creates a predict_abalone_age prediction function, which is used in an SQL command to generate predictions natively in Redshift.

Once the CREATE MODEL statement is completed, you can use the show model command to see the model's status:

```
show model chapter11_byom.predict_abalone_age;
```

Here is the output:

☐ Key	Value
☐ Model Name	predict_abalone_age
☐ Schema Name	public
☐ Owner	IAMR:Admin
☐ Creation Time	Sat, 31.12.2022 17:39:06
☐ Model State	READY
☐	
☐ PARAMETERS:	
☐ Model Type	xgboost
☐ S3 Model Path	s3://redshift-ml-22-redshiftmlbucket-1cckvqgktpfe0/sagemaker/DE…
☐ Function Name	predict_abalone_age
☐ Inference Type	Local
☐ Function Parameter T…	int4 int4 float8 float8 float8 float8 float8 float8 float8
☐ IAM Role	arn:aws:iam::215830370936:role/spectrumrs

Figure 11.1 – Local inference model metadata

Notice that **Model State** is **READY** and **S3 Model Path** is the one we gave when creating the model. **Inference Type** is **Local**, which means the model type is local inference.

We have successfully created the local inference model; now, let's prepare a test dataset to test whether the local inference is working without any issues.

Data preparation

Load the test data from the S3 bucket to a Redshift table to test our local inference model.

> **Note**
>
> Please update IAM_ROLE. Do not change the S3 bucket location.

Run the following command to create the table and load the data:

```
drop table if exists chapter11_byom.abalone_test;

create table chapter11_byom.abalone_test
(Rings int, sex int,Length_ float, Diameter float, Height float,
WholeWeight float, ShuckedWeight float,VisceraWeight float,
```

```
ShellWeight float );
copy chapter11_byom.abalone_test
from 's3://jumpstart-cache-prod-us-east-1/1p-notebooks-datasets/
abalone/text-csv/test/'
IAM_ROLE 'arn:aws:iam::212330312345:role/spectrumrs'
csv ;
```

Sample the test table to make sure the data is loaded:

```
select * from chapter11_byom.abalone_test limit 10;
```

Here is the sample dataset:

☐ rings	sex	length	diameter	height	wholeweight	shuckedw
☐ 11	1	0.585	0.455	0.15	0.987	0.4355
☐ 5	3	0.325	0.245	0.075	0.1495	0.0605
☐ 9	3	0.58	0.42	0.14	0.701	0.3285
☐ 12	2	0.48	0.38	0.145	0.59	0.232
☐ 11	2	0.44	0.355	0.115	0.415	0.1585
☐ 4	3	0.245	0.18	0.065	0.071	0.03
☐ 7	1	0.63	0.51	0.17	1.1885	0.4915
☐ 9	2	0.64	0.505	0.175	1.3185	0.6185
☐ 10	1	0.45	0.335	0.125	0.349	0.119
☐ 11	3	0.56	0.44	0.155	0.811	0.3685

Figure 11.2 – Showing sample records from the test dataset

Now that we have loaded the test data, let's run the SELECT command, which invokes the predict_
abalone_age function.

Inference

Now, call the prediction function that was created as part of the CREATE MODEL statement:

```
Select original_age, predicted_age, original_age-predicted_age as
Error
From(
select predict_abalone_age(Rings,sex,
Length_ ,
Diameter ,
Height ,
WholeWeight ,
ShuckedWeight ,
ViscentWeight ,
ViscaraWeight ,
```

```
ShellWeight ) predicted_age, rings as original_age
from chapter11_byom.abalone_test ) a;
```

Here's the output of the predictions generated using local inference:

original_age	predicted_age	error
11	12	-1
5	6	-1
9	10	-1
12	12	0
11	9	2
4	5	-1
7	11	-4
9	10	-1
10	11	-1
11	10	1
9	10	-1
7	9	-2
7	8	-1

Result 1 (100)

Figure 11.3 – Showing actual versus predicted values

We have successfully trained and validated a SageMaker model and then deployed it to Redshift ML. We also generated predictions using the local inference function. This demonstrates Redshift's BYOM local inference feature.

In the next section, you are going to learn about the BYOM remote inference feature.

BYOM using a SageMaker endpoint for remote inference

In this section, we will explore how to create a BYOM remote inference for an Amazon SageMaker Random Cut Forest model. This means you are bringing your own machine learning model, which is trained on data outside of Redshift, and using it to make predictions on data stored in a Redshift cluster using an endpoint. In this method, to use BYOM for remote inference, a machine learning model is trained, an endpoint is created in Amazon SageMaker, and then the endpoint is accessed from within a Redshift query using SQL functions provided by the Amazon Redshift ML extension.

This method is useful when Redshift ML does not natively support models, for example, a Random Cut Forest model. You can read more about Random Cut Forest here: `https://tinyurl.com/348v8nnw`.

To demonstrate this feature, you will first need to follow the instructions found in this notebook (`https://github.com/aws/amazon-sagemaker-examples/blob/main/introduction_to_amazon_algorithms/random_cut_forest/random_cut_forest.ipynb`) to create a Random Cut Forest machine learning model using Amazon SageMaker to detect anomalies. Please complete the Amazon SageMaker model training and validate the model to make sure the endpoint is working and then proceed to the next section.

Creating BYOM remote inference

Once you have validated that the SageMaker endpoint is deployed and working properly, let's define a `CREATE MODEL` reference point inside Redshift by specifying the SageMaker endpoint. Using the same notebook, let's build the `CREATE MODEL` statement in Jupyter and run it in Query Editor v2.

Setting up the parameters

Let's start by setting up the parameters:

- `S3_Bucket` is used to store Redshift ML artifacts
- `SAGEMAKER_ENDPOINT` is the model endpoint on the SageMaker side to run inferences against
- `REDSHIFT_IAM_ROLE` is the cluster role:

```
REDSHIFT_IAM_ROLE = 'arn:aws:iam::215830312345:role/spectrumrs'
SAGEMAKER_ENDPOINT = rcf_inference.endpoint
```

> **Note**
> Please update `REDSHIFT_IAM_ROLE` with your Redshift cluster role.

Generating the BYOM remote inference command

Let's generate the `CREATE MODEL` statement by running the following code:

```
sql_text=("drop model if exists chapter11_byom.remote_random_cut_forest;\
CREATE MODEL chapter11_byom.remote_random_cut_forest\
 FUNCTION remote_fn_rcf (int)\
 RETURNS decimal(10,6)\
 SAGEMAKER'{}'\
```

```
  IAM_ROLE'{}'\
")
print(sql_text.format(SAGEMAKER_ENDPOINT,REDSHIFT_IAM_ROLE))
```

You have finished the work with the Jupyter notebook. Now you have a pre-trained model in Amazon SageMaker and the next step is to bring it into Redshift ML. To do so, access Query Editor v2, connect to the Serverless endpoint, and run the commands outlined next.

In Query Editor v2, run the following command:

```
CREATE MODEL chapter11_byom.remote_random_cut_forest
FUNCTION remote_fn_rcf (int) RETURNS decimal(10,6)
SAGEMAKER'randomcutforest-2022-12-31-03-48-13-259'
IAM_ROLE'arn:aws:iam:: 215830312345:role/spectrumrs'
;
```

Retrieve the model metadata by running the show model command:

```
show model chapter11_byom.remote_random_cut_forest;
```

The output is as follows:

Key	Value
Model Name	remote_random_cut_forest
Schema Name	public
Owner	IAMR:Admin
Creation Time	Sat, 31.12.2022 18:49:27
Model State	READY
PARAMETERS:	
Endpoint	randomcutforest-2022-12-31-03-48-13-259
Function Name	remote_fn_rcf
Inference Type	Remote
Function Parameter T...	int4
IAM Role	arn:aws:iam::215830370936:role/spectrumrs

Figure 11.4 – Remote inference model metadata

Notice that in the model metadata, the **Model State** parameter is set to **READY**, indicating that the model is deployed. The **Endpoint** name is **randomcutforest-2022-12-31-03-48-13-259**. **Inference Type** is set to **Remote** inference. When this model is run, Redshift ML sends data stored in Redshift in batches to SageMaker, where inferences are generated. Generated predicted values are then sent back to Redshift, which are eventually presented to the user.

We have successfully deployed the model. In the next section, let's run predictions.

The data preparation script

The following code snippet shows the data preparation script that you will need to run on Redshift. We will create the table that will be used to run inference on:

```
COPY chapter11_byom.rcf_taxi_data
FROM 's3://sagemaker-sample-files/datasets/tabular/anomaly_benchmark_
taxi/NAB_nyc_taxi.csv'
IAM_ROLE 'arn:aws:iam::215830312345:role/spectrumrs' ignoreheader 1
csv delimiter ',';
```

> **Note**
>
> Please update the IAM_ROLE parameter with your Redshift cluster attached role.

Sample the data to make sure data is loaded:

```
select * from chapter11_byom.rcf_taxi_data limit 10;
```

Here's the output:

Figure 11.5 – Showing sample records from the test dataset

Now that we have the remote inference endpoint and test dataset, let's invoke the prediction function.

Computing anomaly scores

Now, let's compute the anomaly scores from the entire taxi dataset:

```
select ride_timestamp, nbr_passengers, chapter11_byom.remote_fn_
rcf(nbr_passengers) as score
from chapter11_byom.rcf_taxi_data;
```

The following is the output of the remote inference predictions:

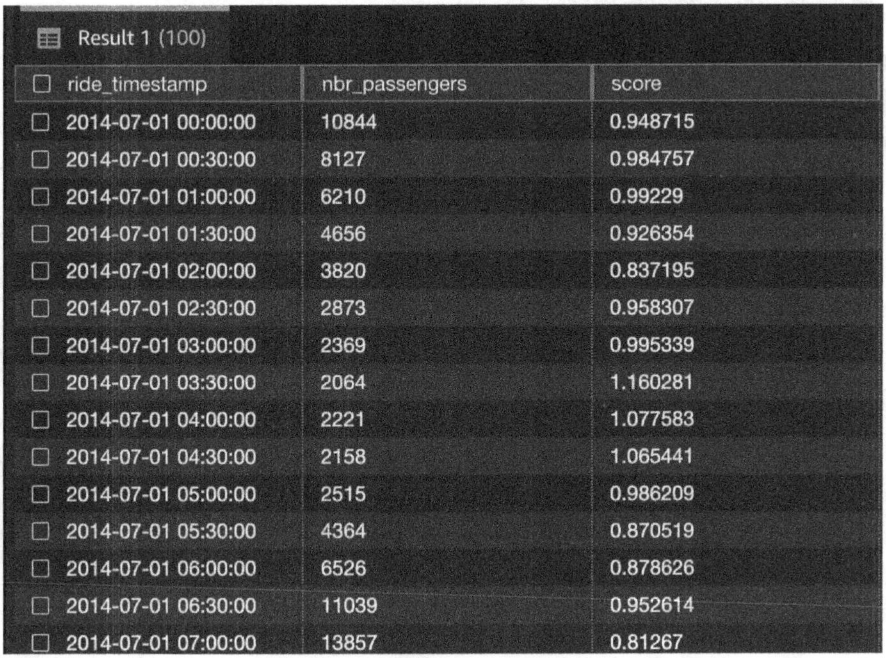

☐ ride_timestamp	nbr_passengers	score
☐ 2014-07-01 00:00:00	10844	0.948715
☐ 2014-07-01 00:30:00	8127	0.984757
☐ 2014-07-01 01:00:00	6210	0.99229
☐ 2014-07-01 01:30:00	4656	0.926354
☐ 2014-07-01 02:00:00	3820	0.837195
☐ 2014-07-01 02:30:00	2873	0.958307
☐ 2014-07-01 03:00:00	2369	0.995339
☐ 2014-07-01 03:30:00	2064	1.160281
☐ 2014-07-01 04:00:00	2221	1.077583
☐ 2014-07-01 04:30:00	2158	1.065441
☐ 2014-07-01 05:00:00	2515	0.986209
☐ 2014-07-01 05:30:00	4364	0.870519
☐ 2014-07-01 06:00:00	6526	0.878626
☐ 2014-07-01 06:30:00	11039	0.952614
☐ 2014-07-01 07:00:00	13857	0.81267

Result 1 (100)

Figure 11.6 – Showing remote function prediction values

The preceding output shows the anomalous score for different days and the number of passengers.

In the following code snippet, we will print any data points with scores greater than 3 and standard deviations (approximately the 99.9th percentile) from the mean score:

```
with score_cutoff as
(select stddev(chapter11_byom.remote_fn_rcf(nbr_passengers)) as std,
avg(chapter11_byom.remote_fn_rcf(nbr_passengers)) as mean, ( mean + 3
* std ) as score_cutoff_value
From chapter11_byom.rcf_taxi_data)
```

```
select ride_timestamp, nbr_passengers, chapter11_byom.remote_fn_
rcf(nbr_passengers) as score
from chapter11_byom.3rcf_taxi_data
where score > (select score_cutoff_value from score_cutoff)
;
```

The output is as follows:

☐ ride_timestamp	nbr_passengers	score
☐ 2014-07-01 18:30:00	27598	2.195869
☐ 2014-07-01 19:00:00	26827	1.830476
☐ 2014-07-02 19:30:00	26872	1.827037
☐ 2014-07-03 19:00:00	29985	3.19337
☐ 2014-07-11 23:30:00	26873	1.847826
☐ 2014-07-14 18:30:00	26945	1.864781
☐ 2014-07-15 19:00:00	27167	1.943336
☐ 2014-07-23 19:00:00	26528	1.726553
☐ 2014-07-23 21:00:00	26600	1.757489
☐ 2014-07-25 23:00:00	26688	1.787407
☐ 2014-09-02 03:30:00	1431	1.992107
☐ 2014-09-05 19:00:00	27337	2.033346
☐ 2014-09-05 19:30:00	26812	1.81532
☐ 2014-09-05 20:00:00	26592	1.730946

Result 1 (100)

Figure 11.7 – Showing unacceptable anomaly scores

In the preceding results, we see that some days' ridership is way higher and our remote inference function is flagging them as anomalous. This concludes the section on bringing remote inference models into Redshift.

Summary

In this chapter, we discussed the benefits and use cases of Amazon Redshift ML BYOM for local and remote inference. We created two SageMaker models and then imported them into Redshift ML as local inference and remote inference model types. We loaded test datasets in Redshift and then we ran the prediction functions and validated both types. This demonstrates how Redshift simplifies and empowers the business community to perform inference on new data using models created outside. This method speeds up the delivery of machine learning models created outside of Redshift to the data warehouse team.

In the next chapter, you are going to learn about Amazon Forecast, which enables you to perform forecasting using Redshift ML.

12

Time-Series Forecasting in Your Data Warehouse

In previous chapters, we discussed how you can use Amazon Redshift **Machine Learning** (ML) to easily create, train, and apply ML models using familiar SQL commands. We talked about how we can use supervised learning algorithms for classification or regression problems to predict a certain outcome. In this chapter, we will talk about how you can use your data in Amazon Redshift to forecast a certain future event using Amazon Forecast.

This chapter will introduce you to time-series forecasting on Amazon Redshift using Amazon Forecast (https://aws.amazon.com/forecast/), a fully managed time-series forecasting service, using SQL, and without moving your data or learning new skills. We will guide you through the following topics:

- Forecasting and time-series data

- What is Amazon Forecast?

- Configuration and security

- Creating forecasting models using Redshift ML

Technical requirements

This chapter requires a web browser and access to the following:

- An AWS account

- Amazon Redshift

- Amazon Redshift query editor v2

You can find the code used in this chapter here: `https://github.com/PacktPublishing/ Serverless-Machine-Learning-with-Amazon-Redshift/blob/main/CodeFiles/ chapter12/chapter-12.sql`.

Forecasting and time-series data

Forecasting is a way of estimating future events, which involves analyzing historical data and past patterns to derive a possible outcome in the future. For example, based on historical data, a business can predict their sales revenue or identify what will happen in the next time period.

Forecasting plays a valuable role in guiding businesses to make informed decisions about their operations and priorities. Many organizations rely on data warehouses such as Amazon Redshift to perform deep analytics on vast amounts of historical and current data, enabling them to drive their business goals and gauge future success. Acting as a planning tool, forecasting helps enterprises prepare for future uncertainties by leveraging past patterns, with the underlying principle that what happened in the past will likely recur in the future. These predictions are based on analyzing observations over time within the given timeframe.

Here are some examples of how organizations use forecasting:

- Financial planning
- Supply and demand planning
- Timing the launch of new products or services
- Resource planning
- Predicting future events, such as sales and revenue earnings
- Reviewing management decisions

Looking at a trend graph helps us predict the trend, but a time-series forecast gives us a better estimate of how it may continue. We can also model data that doesn't show any clear pattern or trend over time. When there is a pattern, we can look at the entire history of the data to see how it happened before. If there is no pattern, we can rely more on recent data for forecasting.

Types of forecasting methods

There are two types of forecasting methods: qualitative and quantitative.

Let's take a look at what qualitative and quantitative methods are, as defined at `https://aws.amazon.com/what-is/forecast/`:

- **Qualitative forecasting** is subjective and relies on marketing experts' opinions to make predictions. You can use these methods when there is not enough historical data. Some examples of qualitative forecasting methods are market research such as polls and surveys, and the Delphi method to collect informed opinions and predict trends.

- **Quantitative forecasting** is objective in nature and is used to predict long-term future trends. It uses historical and current data to forecast future trends. Some examples of quantitative forecasting methods are *time-series forecasting, econometric modeling*, and the *indicator approach*.

In this chapter, we will focus on quantitative forecasting using time series for data, also known as time-series forecasting. Now, let's look into what time-series forecasting is.

What is time-series forecasting?

Time-series forecasting is a data science technique that uses ML to study historical data and predict future trends or behavior in time-series data. Time-series data is used in many situations, such as weather forecasting, financial studies, statistics, resource planning, and econometrics. In the previous chapter, we looked into regression models to predict values using cross-sectional data, where your input variables are used to determine the relationship between the variables so that you can predict the unknown target on sets of data without the target variables.

This data is unique because it arranges data points by time. Time-series data can be plotted on a graph and these graphs are valuable tools for visualizing and analyzing the data. In many organizations, data scientists or data analysts use these graphs to identify forecasting data features or attributes. Let us look into some examples of time-series data characteristics.

Time trending data

In trending data, the observations are captured at equal time intervals. In time-series graphs, the y axis is always a unit of time, such as quarter, year, month, day, hour, minute, or second. In *Figure 12.1*, we have an example of the trend of total subscribers by year:

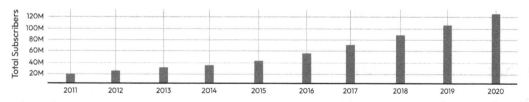

Figure 12.1 – Trend of total subscribers per year

Seasonality

In seasonality observations, we can see periodic fluctuations over time, and these fluctuations are predictable because we understand the behavior and the cause based on historical patterns. For example, retailers know that sales will increase during certain holiday periods. In *Figure 12.2*, we see an upward spike in sales for November and December, which is expected because of the holiday season:

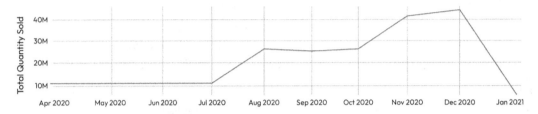

Figure 12.2 – Upward spike due to holiday season

Structural breaks

In structural breaks, we have fluctuations that are less predictable and can occur at any point in time. For example, during a recession or geo-political disturbances, the economic situation of a country might show structural breaks. In *Figure 12.3*, we can see a visualization of economic growth over time. The dips indicate an event that occurred at certain data points; for example, the one in 2009 correlates to the mortgage crisis in the US.

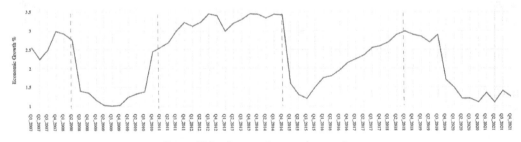

Figure 12.3 – Economic growth over time

Let's take a look into how Amazon Redshift ML uses Amazon Forecast to generate models using time-series datasets.

What is Amazon Forecast?

Amazon Forecast, like Amazon Redshift ML, requires no ML experience to use. Time-series forecasts are generated using various ML and statistical algorithms based on historical data. As a user, you simply send data to Amazon Forecast and it will examine the data and automatically identify what is meaningful and produces a forecasting model.

With Amazon Redshift ML, you can leverage Amazon Forecast to create and train forecasting models from your time-series data and use these models to generate forecasts. For forecasting, we require a target time-series dataset. In target time-series forecasting, we predict the future value of a variable using the past data or previous values, which is often called univariate time series because the data is sequential over equal time increments. Currently, Redshift ML supports target time-series datasets with a custom domain. The dataset in your data warehouse must contain the frequency or interval at which you capture your data. For example, you might record and aggregate the average temperature every hour.

Amazon Forecast automatically trains your model based on an algorithm using Auto ML and provides six built-in algorithms (to learn more about the built-in algorithms, please check out this resource: `https://docs.aws.amazon.com/forecast/latest/dg/aws-forecast-choosing-recipes.html#forecast-algos`). These forecasting models, known as predictors, are created using an optimal combination of these algorithms from your time-series data in Amazon Redshift.

Configuration and security

As Amazon Forecast is a separate fully managed service, you will need to create or modify your IAM role to include access permissions for your serverless endpoint or Redshift cluster. Additionally, you should configure a trust relationship for Amazon Forecast (`forecast.amazonaws.com`) in the IAM role to enable the necessary permissions.

You can use the **AmazonForecastFullAccess** managed policy, which grants full access to Amazon Forecast and all of the supported operations. You can attach this policy to your default role but, in your production environments, you must follow the principle of least-privilege permissions. You may use more restrictive permissions, such as the following:

```
{
    "Version": "2012-10-17",
    "Statement": [
        {
            "Sid": "VisualEditor0",
            "Effect": "Allow",
            "Action": [
                "forecast:DescribeDataset",
                "forecast:DescribeDatasetGroup",
                "forecast:DescribeAutoPredictor",
                "forecast:CreateDatasetImportJob",
                "forecast:CreateForecast",
                "forecast:DescribeForecast",
                "forecast:DescribeForecastExportJob",
                "forecast:CreateMonitor",
                "forecast:CreateForecastExportJob",
                "forecast:CreateAutoPredictor",
```

```
            "forecast:DescribeDatasetImportJob",
            "forecast:CreateDatasetGroup",
            "forecast:CreateDataset",
            "forecast:TagResource",
            "forecast:UpdateDatasetGroup"
        ],
        "Resource": "*"
      },
    {
      "Effect": "Allow",
      "Action": [
        "iam:PassRole"
      ],
      "Resource":"arn:aws:iam::<aws_account_id>:role/service-
role/<Amazon_Redshift_cluster_iam_role_name>"
    }
    ]
}
```

Creating forecasting models using Redshift ML

Currently, if you have to perform forecasting in your data warehouse, you need to export the dataset into external systems and then apply forecasting algorithms to create output datasets and then import them back into the data warehouse for your presentation layer or further analysis. With Redshift ML's integration with Amazon Forecast, you don't have to perform all these steps. You can now create the forecasting models right on your dataset within your data warehouse.

In *Chapter 5*, we talked about the basic CREATE MODEL syntax and its constructs. Let's take a look at the CREATE MODEL syntax for forecasting:

```
CREATE MODEL forecast_model_name
FROM { table_name | ( select_query ) }
TARGET column_name
IAM_ROLE { default | 'arn:aws:iam::<AWS account-id>:role/<role-name>'
}
AUTO ON MODEL_TYPE FORECAST
[ OBJECTIVE optimization_metric ]
SETTINGS (S3_BUCKET 'bucket',
         HORIZON integer,
         FREQUENCY forecast_frequency,
         [, PERCENTILES perc_comma_delim_string],
         [ S3_GARGABE_COLLECT OFF ])
```

There are a few things to notice with the CREATE MODEL statement for forecasting.

First, forecast models do not create inference functions. The reason for this is that when we train a predictor on Amazon Forecast, we specify in the training request the number (HORIZON) and frequency of predictions (FREQUENCY) we want to make in the future. Because of this, a trained model has a fixed forecast, so there isn't a physical model to compile and execute. A custom CTAS command (which will be discussed later) is used to extract a forecast from the training output location in S3 into a table locally in Redshift.

Next, we can specify the *optional* objective or optimization metric, which is used to optimize the predictor for under-forecasting and over-forecasting. Amazon Forecast provides different model accuracy metrics for you to assess the strength of your forecasting models, which are listed here:

- AverageWeightedQuantileLoss – measures the accuracy of a model at a specified quantile
- WAPE (weighted absolute percentage error) – measures the overall deviation of forecasted values from the observed value
- RMSE (root mean square error) – the square root of the average of squared errors
- MASE (mean absolute scaled error) – calculated by dividing the average error by a scaling factor
- MAPE (mean absolute percentage error) – takes the absolute value of the percentage error between observed and predicted values for each unit of time, then averages those values

Lastly, it is important to note that FORECAST does not support any hyperparameters. Instead, any FORECAST-specific settings for training will be specified using the SETTINGS clause. Currently, the supported settings are as follows:

- FREQUENCY: Granularity of predictions in a forecast. Valid values are Y (year), M (month), W (week), D (day), H (hour), and min (minute), for example, H for hourly forecasts or 1min for forecasts every minute).
- HORIZON: The number of time steps in the future to forecast (e.g., 24).

> **Note**
> FREQUENCY H and HORIZON 24 mean you want hourly forecasts for the next day.

- PERCENTILES (optional): The forecast types are used to train a predictor. Up to five forecast types or percentiles can be specified. These types can be quantiles [0.01 to 0.99] or mean. A forecast at the 0.50 quantile will estimate a lower value 50% of the time.

Now, let's take a look at one use case where we can use the target time-series dataset for predicting the target forecast value.

Business problem

For this use case, let's take the example of an online retail store to forecast the future demand for certain products in the store. This dataset is taken from the UCI ML repository and is available here: `https://archive.ics.uci.edu/dataset/352/online+retail`. For this exercise, we have modified the data to resemble more of a target time-series dataset, containing `item_id`, `date`, and `target_value` fields. The data spans a two-year time period starting from December 2018 to November 2020. The modified data contains the item name, date products were sold, and total number of products sold.

> **Dataset citation**
>
> Online Retail. (2015). UCI Machine Learning Repository. `https://doi.org/10.24432/C5BW33`.

Uploading and analyzing the data

After successfully connecting to Redshift as an admin or database developer, load data into Amazon Redshift and follow the steps outlined here:

1. Navigate to query editor v2, connect to **Serverless** endpoint, and connect to the **dev** database:

Figure 12.4 – Connecting to the dev database

2. Execute the following steps to create the schema and the trade details table and load the data:

```
CREATE SCHEMA chapter12_forecasting;

Create table chapter12_forecasting.web_retail_sales
(invoice_Date date, item_id varchar(500), quantity int);

COPY chapter12_forecasting.web_retail_sales
FROM 's3://packt-serverless-ml-redshift/chapter12/web_retail_
sales.csv'
IAM_ROLE default
FORMAT AS CSV
DELIMITER ','
```

```
IGNOREHEADER 1
DATEFORMAT 'YYYY-MM-DD'

REGION AS 'eu-west-1';
```

3. Run the following query to examine some sample data:

    ```
    select * from chapter12_forecasting.web_retail_sales;
    ```

    The result will be similar to this:

invoice_date	item_id	quantity
2019-09-23	JUMBO SHOPPER VINTAGE RED PAISLEY	36
2019-09-30	PACK OF 6 BIRDY GIFT TAGS	67
2019-10-07	REGENCY CAKESTAND 3 TIER	68
2019-10-19	SMALL POPCORN HOLDER	89
2019-12-17	CHARLOTTE BAG SUKI DESIGN	46
2020-02-22	VINTAGE HEADS AND TAILS CARD GAME	4
2020-08-04	6 RIBBONS RUSTIC CHARM	33
2020-09-07	PACK OF 72 RETROSPOT CAKE CASES	50
2020-09-29	HEART OF WICKER LARGE	22
2020-10-25	PACK OF 12 LONDON TISSUES	36
2020-11-06	RED RETROSPOT CHARLOTTE BAG	30

Figure 12.5 – Query results

As you can see in the preceding figure, we have the following:

*   `invoice_date` (date when the item was sold)
*   `item_id` (name of the product sold)
*   `quantity` (number of items sold for that product for each day)

Using this dataset, we will create a model in Amazon Forecast and predict the demand for the future for the given products. The goal is to analyze what a particular product's demand is going to look like in the coming five days. For accuracy and validation, we will create the model using the data until October 2020. Once we have the predictor ready, we will then compare the output values with the actual values in November 2020 to determine the accuracy of our model. We will also take a look at different accuracy metrics, such as the average **weighted quantile loss** (**wQL**), WAPE, MAPE, MASE, and RMSE.

Let's create the model using the CREATE MODEL statement we discussed at the beginning of the *Creating forecasting models using Redshift ML* section.

Objective is set to AverageWeightedQuantileLoss (mean of wQL), which is the accuracy metric for optimization_metric. Frequency is set to D (Days), Horizon is set to 5, and Percentiles is set to 0.25, 0.50, 0.75, 0.90, and mean.

If you do not specify the percentiles settings, then Forecast generates the predictions on p10, p50, and p90 (0.10, 0.50, and 0.90).

Run the following command in query editor v2 to create the model. Note this will take approximately 90 minutes.

```
CREATE MODEL forecast_sales_demand
FROM (select item_id, invoice_date, quantity from  chapter12_
forecasting.web_retail_sales where invoice_date <  '2020-10-31')
TARGET quantity
IAM_ROLE 'arn:aws:your-IAM-Role'
AUTO ON MODEL_TYPE FORECAST
OBJECTIVE 'AverageWeightedQuantileLoss'
SETTINGS (S3_BUCKET '<<bucket name>>',
  HORIZON 5,
  FREQUENCY 'D',
  PERCENTILES '0.25,0.50,0.75,0.90,mean',
  S3_GARBAGE_COLLECT OFF);
```

Run the SHOW MODEL command to see whether model training is complete:

```
SHOW MODEL forecast_sales_demand;
```

The result is as follows:

	Key	Value	≡
☐	Model Name	forecast_sales_demand	
☐	Schema Name	chapter12_forecasting	
☐	Owner	IAMR:Admin	
☐	Creation Time	Sat, 19.08.2023 19:16:38	
☐	Model State	TRAINING	
☐			
☐	TRAINING DATA:		
☐	Query	SELECT ITEM_ID, INVOICE_DATE, QUANTITY	
☐		FROM CHAPTER12_FORECASTING.WEB_RETAIL_SALES	
☐		WHERE INVOICE_DATE < '2020-10-31'	
☐	Target Column	QUANTITY	
☐			
☐	PARAMETERS:		
☐	Model Type	forecast	
☐	Frequency	D	
☐	Horizon	5	
☐	Percentiles	0.25, 0.50, 0.75, 0.90, mean	
☐	Optimization Metric	AverageWeightedQuantileLoss	
☐	Training Job Name	redshiftml_20230819191638640405	

Figure 12.6 – Result of model training

You can also view the status of the predictor using the value of **Training Job Name** shown in the preceding screenshot. Navigate to your AWS console and type in Amazon Forecast.

Click on **View dataset groups** and find the dataset group name by pasting **redshiftml_20221224001451333090**.

Click on this dataset group name and verify whether **Target time series data** is **Active**, as shown in *Figure 12.7*.

You can also view the details about your time-series data by clicking **View** and seeing the schema, frequency of data registered in your data file, dataset import details, and so on.

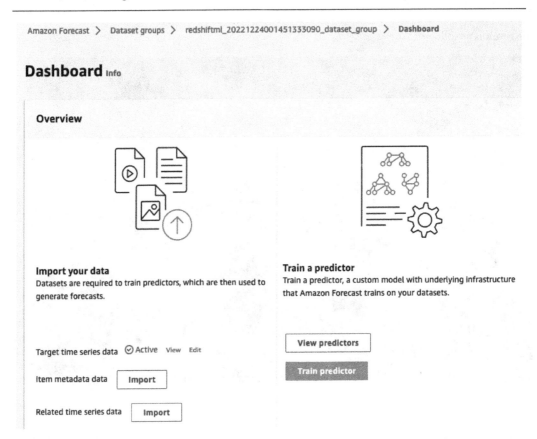

Figure 12.7 – Verify the status of target time-series data

Once active, you can view the predictor by clicking **View predictors**. The **Predictors** dialog box will show the training status, as follows:

Figure 12.8 – Training status in View predictors

Run the SHOW MODEL command again to see if model training is complete:

```
SHOW MODEL forecast_sales_demand;
```

The result is as follows:

☐	Model Name	forecast_sales_demand
☐	Schema Name	chapter12_forecasting
☐	Owner	IAMR:Admin
☐	Creation Time	Sat, 19.08.2023 19:16:38
☐	Model State	READY
☐		
☐	TRAINING DATA:	
☐	Query	SELECT ITEM_ID, INVOICE_DATE, QUANTITY
☐		FROM CHAPTER12_FORECASTING.WEB_RETAIL_SALES
☐		WHERE INVOICE_DATE < '2020-10-31'
☐	Target Column	QUANTITY
☐		
☐	PARAMETERS:	
☐	Model Type	forecast
☐	Frequency	D
☐	Horizon	5
☐	Percentiles	0.25, 0.50, 0.75, 0.90, mean
☐	Optimization Metric	AverageWeightedQuantileLoss
☐	Training Job Name	redshiftml_20230819191638640405

Figure 12.9 – Status of model training completion

Once the model training is finished and ready, you can then view the outputs by creating a table on your forecast:

For a retail store, the company needs to ensure that they do not over-forecast or under-forecast the predicted quantity required in order to effectively manage inventory and enhance profits. As mentioned earlier, Redshift ML with Amazon Forecast provides different optimization metrics that can be used in order to measure the accuracy of a model specified at different quantiles. For this use case, we have created the model for 0.25, 0.50, 0.75, 0.90, and mean. If the emphasis is on over-forecasting, then for a retailer, choosing a higher quantile (0.90) captures the spike in demand in a much better way for a high-demand item or product. This suggests that there is a 90% probability of success for the product to meet the forecasted demand. Now, let's see how to get our forecasted results.

Creating a table with output results

After the model has finished training and is ready, we now create a table in our schema to hold all the forecast results using a simple CTAS command, as shown:

```
create table chapter12_forecasting.tbl_forecast_sales_demand as SELECT
FORECAST(forecast_sales_demand);
```

In this command, `forecast ()` is a pseudo table function that takes the name of your model as an input parameter. The data is then pulled from the S3 bucket location where your model results are stored.

Let's take a look at the output from the preceding table by running the following SQL command:

```
select * from chapter12_forecasting.tbl_forecast_sales_demand;
```

Looking at *Figure 12.10*, you can see that for each day, Forecast has generated the output predictions for each distribution point or quantile that we provided and the mean:

id	time	p25	p50	p75	p90	mean
white hanging heart t-light holder	2020-11-01T00:00:00Z	20.2240104675	46.8012008667	71.7090988159	112.0020904541	44.9173049927
white hanging heart t-light holder	2020-11-02T00:00:00Z	-2.0889434814	34.4102172852	66.7431182861	97.9138565063	36.0352134705
white hanging heart t-light holder	2020-11-03T00:00:00Z	23.2653865814	51.943359375	78.2389526367	113.6890335083	49.3352928162
white hanging heart t-light holder	2020-11-04T00:00:00Z	17.1822776794	43.1986045837	70.4421844482	113.2993164062	37.2828941345
white hanging heart t-light holder	2020-11-05T00:00:00Z	-4.0084266663	18.7009906769	43.2981262207	61.9549255371	13.0443763733

Figure 12.10 – Output of the table

For products in high demand, the retailer can choose a higher quantile, such as 0.90 (p90), which better captures spikes in demand, rather than forecasting at the mean or 0.50 quantile.

Now, let's take a look at the data of a popular product: **JUMBO BAG RED RETROSPOT**.

Run the following SQL query:

```
select a.item_id as product,
a.invoice_date,
a.quantity as actual_quantity ,
p90::int as p90_forecast,
p90::int - a.quantity as p90_error ,mean::int,
p50::int as p50_forecast
from chapter12_forecasting.web_retail_sales a
inner join chapter12_forecasting.tbl_forecast_sales_demand b
on upper(a.item_id) = upper(b.id)
and a.invoice_date = to_date(b.time, 'YYYY-MM-DD')
AND a.item_id = 'JUMBO BAG RED RETROSPOT'
where invoice_date > '2020-10-31'
order by 1,2;
```

Here's the result:

product	invoice_date	actual_quantity	p90_forecast	p90_error	mean	p50_forecast
JUMBO BAG RED RETR...	2020-11-01	318	415	97	181	181
JUMBO BAG RED RETR...	2020-11-02	244	222	-22	86	86
JUMBO BAG RED RETR...	2020-11-03	452	334	-118	131	131
JUMBO BAG RED RETR...	2020-11-04	369	548	179	166	166

Figure 12.11 – Forecast data

To visualize the data, select **Chart**. For the *x* axis, choose the `invoice_date` attribute, and for the *y* axis, choose `p90_forecast`:

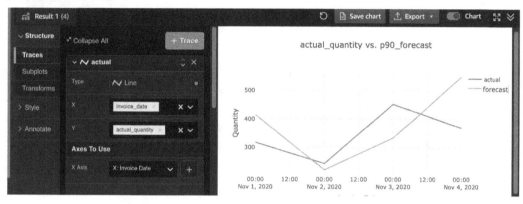

Figure 12.12 – Forecast chart

If we closely examine the preceding data in *Figure 12.11*, we can observe that Line 1 was under-forecasted, while Lines 2 and 3 were very close to the actual values, and Line 4 was just slightly over-forecasted. In order to test the forecasting, you can further perform tests with different sets of data or even on different quantiles. Additionally, a retailer can use this data for different products, such as products with low demand, and make use of other quantiles, such as p50 or `mean`.

The wQL is used to calculate the `AverageWeightedQuantileLoss` metric. The wQL can be used to manage the costs of over- and under-forecasting. These metrics will be available to you in the Amazon Forecast console for your predictor. Generally, to calculate the wQL at 0.90, sum the values of the positive values in above p90 error field and multiply them by a smaller weight of 0.10, and sum the absolute values of the negative values in p90 error and multiply them by 0.90.

To align with your business outcomes, you can create the forecasting models at different quantiles (**Percentiles**) in your Amazon Redshift data warehouse. This gives you the flexibility to measure your business goals and keep the impacts on cost on the lower side.

Summary

In this chapter, we discussed how you can use Redshift ML to generate forecasting models using Amazon Forecast by creating the model for Forecast `Model_Type`. You learned about what forecasting is and how time-series data is used to generate different models for different quantiles. We also looked at different quantiles and talked briefly about different optimization metrics.

We showed how forecast models can be used to predict the future quantity sale for a retailer use case and how they can be used to balance the effect of over-forecasting and under-forecasting.

In the next chapter, we will look at operational and optimization considerations.

13

Operationalizing and Optimizing Amazon Redshift ML Models

Now that you have learned how to create many different types of ML models, we will show you how you can operationalize your model training pipelines. Once you have moved your model to production, you want to refresh the model regularly and automate the process to do this. Additionally, it is important to periodically evaluate your models to maintain and improve their accuracy.

In this chapter, we will go through the following main topics:

- Operationalizing your ML models
- Optimizing the Redshift model for accuracy

Technical requirements

This chapter requires a web browser and access to the following:

- An AWS account
- An Amazon Redshift Serverless endpoint
- Amazon Redshift Query Editor v2
- An Amazon EC2 Linux instance (optional)

You can find the code used in this chapter here: `https://github.com/PacktPublishing/Serverless-Machine-Learning-with-Amazon-Redshift/`.

Operationalizing your ML models

Once a model is validated and used on a regular basis for running predictions, it should be operationalized. The reasons for this are to remove the manual tasks of retraining your models and to ensure that your model still retains high accuracy after your data distribution has changed over time, also referred to as **data drift**. When data drift occurs, you need to retrain the model using an updated training set.

In the following sections, we will do a simple model retraining, then show you how you can create a version from an existing model.

Model retraining process without versioning

To walk through the retraining process, we will use one of our previously used models.

In *Chapter 7*, we discussed different regression models, so let's use the `chapter7_regressionmodel.predict_ticket_price_auto` model. This model solved a multi-input regression problem and **SageMaker Autopilot** chose the **XGBoost algorithm**.

Let's assume this model is performing well and, based on our data loading processes, we want to retrain this model weekly.

To retrain this model, we must first remove the existing model and then re-execute the `CREATE MODEL` command as follows:

```
DROP MODEL chapter7_RegressionMOdel.predict_ticket_price_auto;
CREATE MODEL chapter7_RegressionMOdel.predict_ticket_price_auto from
chapter7_RegressionModel.sporting_event_ticket_info_training
TARGET final_ticket_price
FUNCTION predict_ticket_price_auto
IAM_ROLE default
PROBLEM_TYPE regression
OBJECTIVE 'mse'
SETTINGS (s3_bucket <<'your-S3-bucket>>',
s3_garbage_collect off,
max_runtime 9600);
```

You can set this up to run on a regular schedule using various techniques, which could include using the Query Editor v2 scheduling feature or running scripts. For more information on scheduling queries with Query Editor v2, refer to the following:

https://docs.aws.amazon.com/redshift/latest/mgmt/query-editor-v2-schedule-query.html.

The model retraining process with versioning

This approach of simply dropping and recreating the model might be fine in some cases, but there is no model history available since we are simply dropping and recreating the model. This makes comparing the newly trained model to previous versions very difficult, if not impossible.

At the time of writing, Redshift ML does not have native versioning capabilities. However, you can still do versioning by implementing a few simple SQL techniques and leveraging the **bring our own model (BYOM)** capability, which you learned about in *Chapter 11*.

BYOM is great for leveraging pre-built Amazon SageMaker models in order to run your inference queries in Amazon Redshift and you can also use BYOM for models that were built using Redshift ML, which means we can create a *version* of an existing model that was previously created by Redshift ML.

Let's take a quick refresher on the syntax of BYOM for local inference:

```
CREATE MODEL model_name
    FROM ('job_name' | 's3_path' )
    FUNCTION function_name ( data_type [, ...] )
    RETURNS data_type
    IAM_ROLE { default }
    [ SETTINGS (
       S3_BUCKET 'bucket', | --required
       KMS_KEY_ID 'kms_string') --optional
    ];
```

We need the job name, the data types of the model inputs, and the output. We can get the information we need for the CREATE MODEL statement by running the SHOW MODEL statement on our existing model. Run the following command in Query Editor v2:

```
SHOW MODEL chapter7_regressionmodel.predict_ticket_price_auto;
```

The result is as follows:

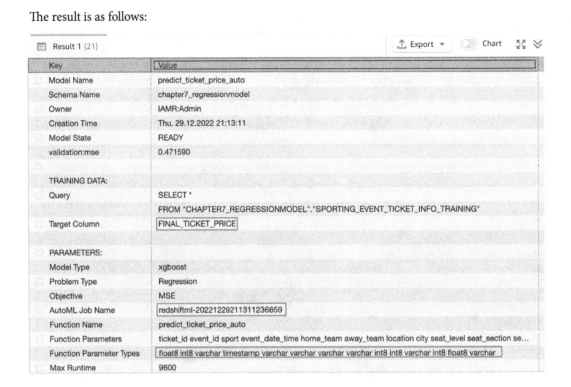

Figure 13.1 – The SHOW MODEL output

The following is the CREATE MODEL statement to create a version of the current model using the **AutoML Job Name** value from our SHOW MODEL command. You will also need to include the function parameter types from *Figure 13.1* in FUNCTION here and include the data type of Target Column(FINAL_TICKET_PRICE). Note that we append the date (YYYYMMDD) to the end of the model name and function name to create our version. You can run the following code in Query Editor v2 to create a version of your model:

```
CREATE MODEL chapter7_regressionmodel.predict_ticket_price_
auto_20230624
    FROM 'redshiftml-20221229211311236659'
    FUNCTION predict_ticket_price_auto_20230624 (float8,
        int8, varchar, timestamp, varchar, varchar,
        varchar, varchar, int8, int8, varchar, int8,
        float8, varchar)
    RETURNS float8
    IAM_ROLE default
    SETTINGS (
      S3_BUCKET '<<your S3 Bucket>>');
```

Run the following SHOW MODEL command:

```
SHOW MODEL chapter7_regressionmodel.predict_ticket_price_
auto_20230624;
```

In *Figure 13.2*, notice that **Inference Type** shows **Local**, which designates this as BYOM with local inference:

Key	Value	
Model Name	predict_ticket_price_auto_20230624	
Schema Name	chapter7_regressionmodel	
Owner	IAMR:Admin	
Creation Time	Sat, 24.06.2023 14:10:35	
Model State	READY	
PARAMETERS:		
Model Type	xgboost	
AutoML Job Name	redshiftml-20230511225221235425	
Function Name	predict_ticket_price_auto_20230624	
Inference Type	Local	
Function Parameter Types	float8 int8 varchar timestamp varchar varchar varchar varchar int8 int8 varchar int8 fl…	
IAM Role	arn:aws:iam::970580199101:role/service-role/AmazonRedshift-CommandsAccessRo…	

Figure 13.2 – The SHOW MODEL output

Now that you have learned how to create a version of a previously trained Redshift ML model, we will show you how you can automate this process.

Automating the CREATE MODEL statement for versioning

We have included the scripts here: https://github.com/PacktPublishing/Serverless-Machine-Learning-with-Amazon-Redshift/tree/main/CodeFiles/chapter13.

You can use these scripts and customize them as needed. These contain all the components needed to automate the process of performing model versioning. The example in this chapter uses Bash scripts with RSQL running on an EC2 instance. If you prefer, you can also install RSQL on Windows or macOS.

You may find more information on using RSQL to interact with Amazon Redshift here: https://docs.aws.amazon.com/redshift/latest/mgmt/rsql-query-tool-getting-started.html.

To download all the code for this book, you may run the commands given in the following link on an EC2 instance running on Linux or Windows or on your local Windows or Mac machine:

`https://github.com/PacktPublishing/Serverless-Machine-Learning-with-Amazon-Redshift.git`.

Before running the scripts, we need to create the schema and the table needed to generate the CREATE MODEL command for the model version. You can run the following steps in Query Editor v2:

1. Create the schema:

    ```
    Create schema chapter13;
    ```

2. Create the table to contain the metadata needed to auto-generate the CREATE MODEL command:

    ```
    create table chapter13.local_inf_ml_model_components
    (model_name varchar(500),
    schema_name varchar(500),
    automlJobName varchar(500),
    functionName varchar(500),
    inputs_data_type varchar(500),
    target_column varchar(50),
    returns_data_type varchar(50),
    model_arn varchar (500),
    S3_Bucket varchar (200) );
    ```

3. Initialize the `local_inf_ml_components` table.

    Note that you will just need to initialize this table once, with the model name, schema name, the data type of the target value we are predicting, the **Amazon Resource Name (ARN)** of the IAM role, and the S3 bucket to be used for the Redshift ML artifacts. The table will get updated with the additional data needed as part of the automation script:

    ```
    insert into chapter13.local_inf_ml_model_components
    values
    (
    'predict_ticket_price_auto',
    'chapter7_regressionmodel',
    ' ',' ',' ',' ','float8',
    '<arn of your IAM ROLE>'
    '<your S3 Bucket>)';
    ```

Now, we are ready to run the automation script. *Figure 13.3* illustrates this flow using our `predict_ticket_price_auto` model from *Chapter 7*. **Step 1** creates the model version by using BYOM and appending the timestamp and **Step 2** drops and creates the new model:

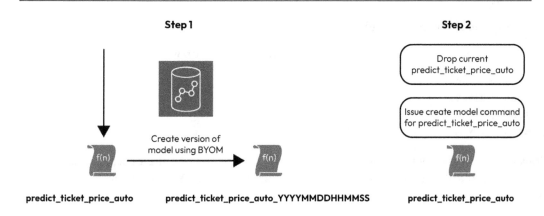

Figure 13.3 – Automation script steps 1 and 2

Let's walk through the steps in *Figure 13.3*.

Step 1 – creating a version from the existing model

You may refer to the step1_create_model_version.sh script at https://github.com/
PacktPublishing/Serverless-Machine-Learning-with-Amazon-Redshift/
tree/main/CodeFiles/chapter13 or where you placed the file after running the git
clone command.

The contents of the step1_create_model_version.sh script are also shown in the following
code snippet. As you can see, it calls other scripts and commands as follows:

```
#! /bin/bash
# create SHOW MODEL sql command
./generate_show_model_sql.sh  'chapter7_regressionmodel.predict_
ticket_price_auto'
#Read SHOW MODEL output and write to file
./show_model.sh
#copy SHOW MODEL output to the model info table
aws s3 cp create_model.txt s3://<your-s3-bucket>>
#load SHOW MODEL output and prep table to generate create model
./prep_create_model.sh
#generate sql to create model version
./generate_create_model_version_sql.sh
#execute the sql to create model verson
./execute_create_model_version.sh
```

Before you execute this script, read through the following subsections as they contain instructions
on some setup steps.

Creating the show_model_sql command

We have a simple script called `generate_show_model_sql.sh` with code as shown here:

```bash
#!/bin/bash
modelname=$1
echo $1
echo SHOW MODEL $1 ';' > show_model.sql
```

This script takes as input the model name. In the script provided, we have already supplied the model name in the `step1_create_model_version.sh` driver script. You can modify this as needed for your models.

The script creates a `SHOW MODEL` command that is written to a file called `show_model.sql` to be read in the `show_model.sh` script.

Reading the SHOW MODEL output and writing it to a file

This step executes an Amazon Redshift RSQL script called `show_model.sh`, which reads the `show_model.sql` file and writes the output to a file called `create_model.txt`.

Copying the SHOW MODEL output to the model info table

This copies the `create_model.txt` file into an S3 bucket.

Loading the SHOW MODEL output and prepping the table to generate CREATE MODEL

This step executes another Amazon Redshift RSQL script called `prep_create_model.sh`, which performs the following:

- Creates and loads the `model_info` table
- Updates `local_inf_ml_model_components` from the `model_info` table so that the `CREATE MODEL` statement can be generated for the model version
- Inserts the generated `CREATE MODEL` statement into the `create_model_sql` table

Generating the SQL to create the model version

This step calls an Amazon Redshift RSQL script called `generate_create_model_version_sql.sh`, which reads the `create_model` table and writes the SQL to a text file called `model_version.txt`.

Executing the SQL to create the model version

This step calls an Amazon Redshift RSQL script called `execute_create_model_version.sh`, which creates the version of our previously created model.

Now you can drop and create your model since we have the model version.

Step 2 – retraining your Redshift ML model to create a version from the existing model

This step calls an Amazon Redshift RSQL script called `retrain_model.sh`, which drops and creates our model. It references `retrain_model.sql`, which you can modify for your needs.

Now that you have learned how to automate the process of retraining your Redshift ML models, let's discuss how to optimize the accuracy of your models.

Optimizing the Redshift models' accuracy

In this section, we will review best practices for maintaining the optimal accuracy of your models.

You will need to continually monitor your models over time to ensure the scores stay stable between model training runs. Consider the new version of the model we created here:

Result 1 (24)

Key	Value
Model Name	predict_ticket_price_auto_new
Schema Name	chapter7_regressionmodel
Owner	admin
Creation Time	Thu, 06.04.2023 10:53:07
Model State	READY
validation:mse	0.424510
Estimated Cost	6.466194
TRAINING DATA:	
Query	SELECT *
	FROM "CHAPTER7_REGRESSIONMODEL"."SPORTING_EVENT_TICKET...
Target Column	FINAL_TICKET_PRICE
PARAMETERS:	

Figure 13.4 – New model output

Create a table similar to this and track each week's mean square error (MSE) score from the SHOW MODEL output:

```
CREATE TABLE chapter13.model_score_history (
    model_name character varying(500),
    schema_name character varying(500),
    score integer,
    variance integer,
    training_date date
)
DISTSTYLE AUTO;
```

The variance will be the difference in the score of each successive version of a model.

Check how your models are trending by writing a query like this:

```
Select model_name, score, variance, training_date
Order by model_name, training_date desc;
```

If variances are not within a reasonable amount, you will need to look at ways to improve the model scores.

Let's explore how we can improve the model quality by using more data and experimenting with different model types and algorithms.

Model quality

The first best practice is to use more data to improve the model's quality. Also, you can add more training time to your model by increasing the MAX_RUNTIME parameter.

Ensure you are using a representative dataset for training and create at least a 10% sample for validation.

Experiment with different model types and algorithms to get the best model. For example, in *Chapter 7*, we tried two different algorithms for the multi-input regression models. On the first one, we tried linear learning and we got an MSE score of 701:

Key	Value
Model Name	predict_ticket_price_linlearn
Schema Name	public
Owner	admin
Creation Time	Tue, 02.08.2022 02:01:58
Model State	READY
validation:mse	701.492249
Estimated Cost	6.687761
TRAINING DATA:	
Query	SELECT *
	FROM "PUBLIC"."SPORTING_EVENT_TICKET_INFO_TRAINING"
Target Column	FINAL_TICKET_PRICE
PARAMETERS:	
Model Type	linear_learner
Problem Type	Regression
Objective	MSE
AutoML Job Name	redshiftml-20220802020158515015
Function Name	predict_ticket_price_linlearn
Function Parameters	ticket_id event_id sport event_date_time home_team away_tea…

Result 1 (24)

Figure 13.5 – MSE score of the linear learner model type

When we ran it again without specifying the model type, SageMaker Autopilot chose XGBoost as the model type and it gave a better MSE score of **0.711260**:

Key	Value	≡
Model Name	predict_ticket_price_auto	
Schema Name	public	
Owner	admin	
Creation Time	Sat, 30.07.2022 18:30:10	
Model State	READY	
validation:mse	0.711260	
Estimated Cost	6.976794	
TRAINING DATA:		
Query	SELECT *	
	FROM "PUBLIC"."SPORTING_EVENT_TICKET_INFO_TRAINI...	
Target Column	FINAL_TICKET_PRICE	
PARAMETERS:		
Model Type	xgboost	
Problem Type	Regression	
Objective	MSE	
AutoML Job Name	redshiftml-20220730183010760167	

Figure 13.6 – MSE score of XGBoost model type

Model explainability

The second best practice is to use the explainability report to better understand which inputs to your model carried the most weight.

Run the following SQL command in Query Editor v2:

```
select EXPLAIN_MODEL ('chapter7_regressionmodel.predict_ticket_price_
auto')
```

This returns Shapley values for the inputs used to train the model:

```
{"explanations":{"kernel_shap":{"label0":{"expected_
value":23.878915786743165,"global_shap_values":
{"away_team":0.050692683450484,"city":0.004979335962039937,"event_
date_time":0.05925819534780525,"event_id":0.31961543069587136,"home_
team":0.04245607437910639,"list_ticket_
price":36.364129559427869,"location":0.005178670063000977,
"seat":0.011496876723927165,"seat_level":0.011342097571256795,
"seat_row":0.011987498536296578,"seat_section":12.15498245617505,
"sport":0.0029737602051575346,"ticket_id":0.3184045531012407,
```

```
"ticketholder":0.005226471657467846}}}},
"version":"1.0"}
```

You will notice that `list_ticket_price` has the highest value of `36.364` – this means it was the highest weighted input. You can experiment by removing the input columns with very low weights as inputs to your model training. Check to see whether you still get the same approximate model score by removing unnecessary columns for the training input and helping improve training times.

Probabilities

For classification problems, leverage the built-in function that is generated so that you can see the probability of a given prediction. Refer to *Chapter 5* for detailed examples of this.

Let's now take a look at some useful notebooks that are generated by Amazon SageMaker Autopilot.

Using SageMaker Autopilot notebooks

Your Autopilot job generates a data exploration notebook and a candidate definition notebook. To view these notebooks, follow these steps:

1. In the AWS console, search for `SageMaker`, then choose **Amazon SageMaker**:

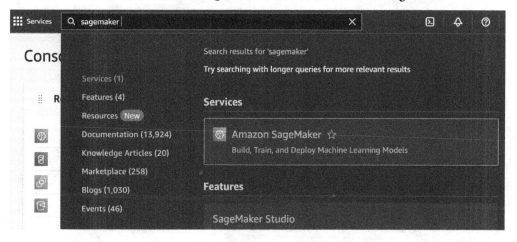

Figure 13.7 – Choosing Amazon SageMaker

2. Then, choose **Studio**:

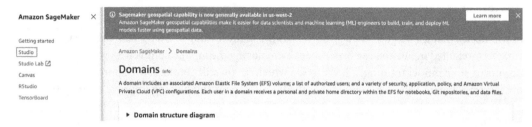

Figure 13.8 – Choosing Studio

3. Next, choose **Open Studio**:

Figure 13.9 – Choosing Open Studio

4. Next, choose **AutoML**:

Figure 13.10 – Choosing AutoML

After you choose **AutoML**, the following screen will show up:

Figure 13.11 – List of model names

5. Choose the model name you want to evaluate. You can get this by using the AutoML job name from your SHOW MODEL output. In this example, I used SHOW MODEL on the predict_ ticket_price_auto model:

Key	Value	
Model Name	predict_ticket_price_auto	
Schema Name	chapter7_regressionmodel	
Owner	IAMR:Admin	
Creation Time	Wed, 15.02.2023 17:46:38	
Model State	READY	
validation:mse	0.800670	
Estimated Cost	8.401197	
TRAINING DATA:		
Query	SELECT *	
	FROM "CHAPTER7_REGRESSIONMODEL"."SPORTING...	
Target Column	FINAL_TICKET_PRICE	
PARAMETERS:		
Model Type	xgboost	
Problem Type	Regression	
Objective	MSE	
AutoML Job Name	redshiftml-20230215174638030561	

Figure 13.12 – The SHOW MODEL output

You will see output like this:

Figure 13.13 – AutoML best model

In *Figure 13.13*, you can see a list of models that were trained, and the *best* model is highlighted. This also shows the objective of **Mse**, the values, and which algorithm was used, and there are links to view the model details, the candidate generation notebook, and the data exploration notebook.

6.   Click on **View model details** – this is another way you can see feature importance or explainability:

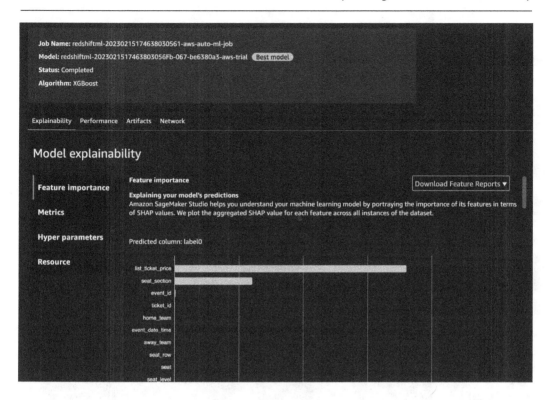

Figure 13.14 – Feature importance

You can also see the hyperparameters used by SageMaker Autopilot:

Figure 13.15 – Hyperparameters

7.   Now, try clicking on **Open data exploration notebook**:

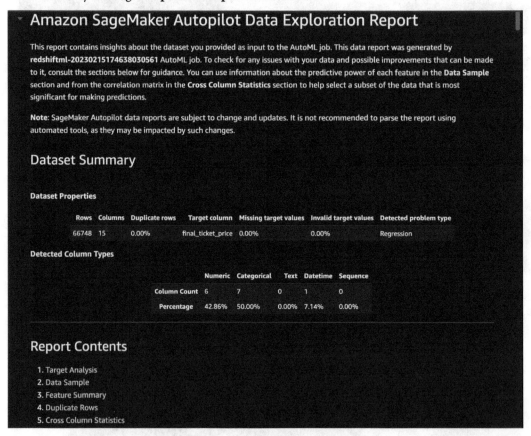

Figure 13.16 – Data exploration report

This will show you the data exploration report and you can see things such as **Target Analysis**, **Feature Summary**, **Duplicate Rows**, and other statistics.

Here is what **Target Analysis** showed for our `predict_ticket_price_auto` model:

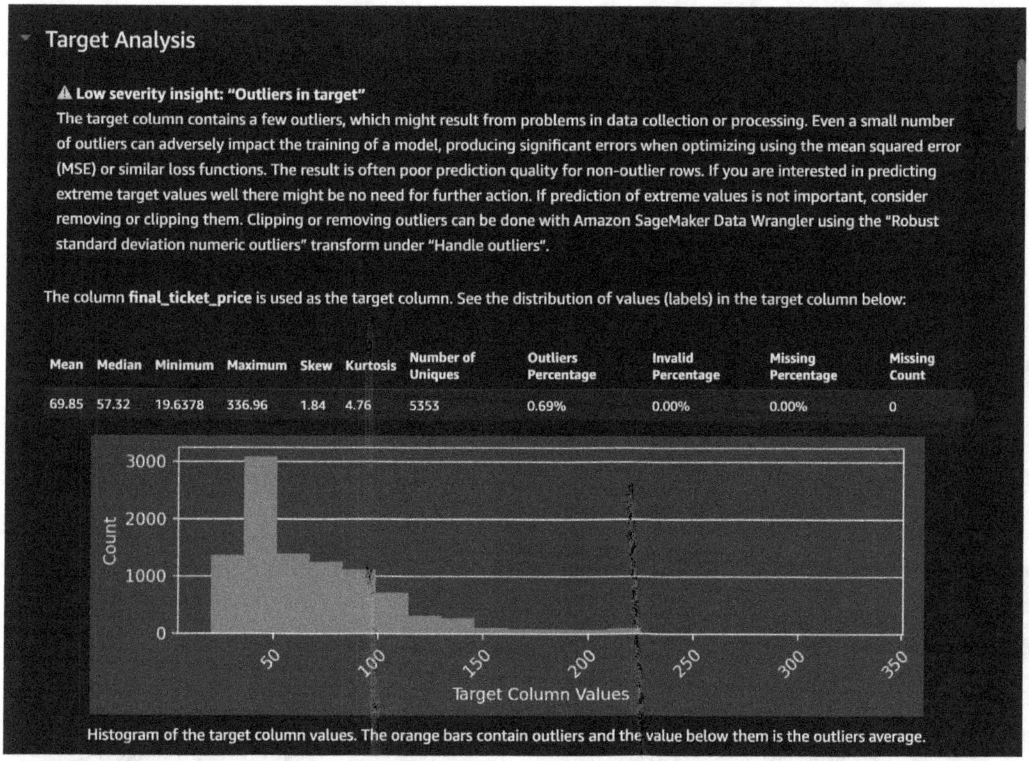
Target Analysis
Low severity insight: "Outliers in target"
The target column contains a few outliers, which might result from problems in data collection or processing. Even a small number of outliers can adversely impact the training of a model, producing significant errors when optimizing using the mean squared error (MSE) or similar loss functions. The result is often poor prediction quality for non-outlier rows. If you are interested in predicting extreme target values well there might be no need for further action. If prediction of extreme values is not important, consider removing or clipping them. Clipping or removing outliers can be done with Amazon SageMaker Data Wrangler using the "Robust standard deviation numeric outliers" transform under "Handle outliers".
The column final_ticket_price is used as the target column. See the distribution of values (labels) in the target column below:
Mean | Median | Minimum | Maximum | Skew | Kurtosis | Number of Uniques | Outliers Percentage | Invalid Percentage | Missing Percentage | Missing Count
69.85 | 57.32 | 19.6378 | 336.96 | 1.84 | 4.76 | 5353 | 0.69% | 0.00% | 0.00% | 0
3000
2000
1000
0
Count
50
100
150
200
250
300
350
Target Column Values
Histogram of the target column values. The orange bars contain outliers and the value below them is the outliers average.

Figure 13.17 – Target Analysis

To learn more about the data exploration notebook, you may refer to this link: `https://docs.aws.amazon.com/sagemaker/latest/dg/autopilot-data-exploration-report.html`.

8.   Now, click on **Open candidate generation notebook**:

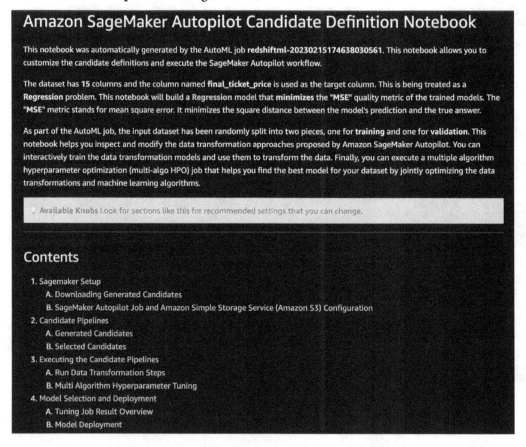

Figure 13.18 – Candidate definition notebook

This notebook contains information about the processing steps, algorithms, and hyperparameters. To learn more about using the candidate generation notebook, refer to `https://docs.aws.amazon.com/sagemaker/latest/dg/autopilot-candidate-generation-notebook.html`.

Summary

In this chapter, you learned techniques to operationalize your models in Amazon Redshift ML.

We discussed how you can create a version of your model. This is important to track the quality of your model over time and to be able to run inferences with different versions.

We then showed you how to optimize your Redshift ML models for accuracy and how you can use the notebooks generated by Amazon SageMaker Autopilot to deepen your understanding of tasks that Autopilot is performing.

We hope you have found this book useful. Our goal when we set out to write this book was to help you gain confidence in these main areas:

- Gaining a better understanding of machine learning and how to use it to solve everyday business problems
- Implementing an end-to-end serverless architecture for ingestion, analytics, and machine learning using Redshift Serverless and Redshift ML
- Creating supervised and unsupervised models, and various techniques to influence your model
- Running inference queries at scale in Redshift to solve a variety of business problems using models created with Redshift ML or natively in Amazon SageMaker

We thank you very much for your time and investment in reading this book. We would welcome your feedback on how we can make Redshift and Redshift ML better. You can find us on LinkedIn.

Index

X

Packtpub.com

Subscribe to our online digital library for full access to over 7,000 books and videos, as well as industry leading tools to help you plan your personal development and advance your career. For more information, please visit our website.

Why subscribe?

- Spend less time learning and more time coding with practical eBooks and Videos from over 4,000 industry professionals

- Improve your learning with Skill Plans built especially for you

- Get a free eBook or video every month

- Fully searchable for easy access to vital information

- Copy and paste, print, and bookmark content

Did you know that Packt offers eBook versions of every book published, with PDF and ePub files available? You can upgrade to the eBook version at packtpub.com and as a print book customer, you are entitled to a discount on the eBook copy. Get in touch with us at customercare@packtpub.com for more details.

At www.packtpub.com, you can also read a collection of free technical articles, sign up for a range of free newsletters, and receive exclusive discounts and offers on Packt books and eBooks.

Other Books You May Enjoy

If you enjoyed this book, you may be interested in these other books by Packt:

Enhancing Deep Learning with Bayesian Inference

Matt Benatan, Jochem Gietema, Marian Schneider

ISBN: 9781803246888

- Understand advantages and disadvantages of Bayesian inference and deep learning
- Understand the fundamentals of Bayesian Neural Networks
- Understand the differences between key BNN implementations/approximations
- Understand the advantages of probabilistic DNNs in production contexts
- How to implement a variety of BDL methods in Python code
- How to apply BDL methods to real-world problems
- Understand how to evaluate BDL methods and choose the best method for a given task
- Learn how to deal with unexpected data in real-world deep learning applications

Natural Language Understanding with Python

Deborah A. Dahl

ISBN: 9781804613429

- Explore the uses and applications of different NLP techniques
- Understand practical data acquisition and system evaluation workflows
- Build cutting-edge and practical NLP applications to solve problems
- Master NLP development from selecting an application to deployment
- Optimize NLP application maintenance after deployment
- Build a strong foundation in neural networks and deep learning for NLU

Packt is searching for authors like you

If you're interested in becoming an author for Packt, please visit authors.packtpub.com and apply today. We have worked with thousands of developers and tech professionals, just like you, to help them share their insight with the global tech community. You can make a general application, apply for a specific hot topic that we are recruiting an author for, or submit your own idea.

Share Your Thoughts

Now you've finished *Serverless Machine Learning with Amazon Redshift ML*, we'd love to hear your thoughts! Scan the QR code below to go straight to the Amazon review page for this book and share your feedback or leave a review on the site that you purchased it from.

https://packt.link/r/1-804-61928-0

Your review is important to us and the tech community and will help us make sure we're delivering excellent quality content.

Download a free PDF copy of this book

Thanks for purchasing this book!

Do you like to read on the go but are unable to carry your print books everywhere? Is your eBook purchase not compatible with the device of your choice?

Don't worry, now with every Packt book you get a DRM-free PDF version of that book at no cost.

Read anywhere, any place, on any device. Search, copy, and paste code from your favorite technical books directly into your application.

The perks don't stop there, you can get exclusive access to discounts, newsletters, and great free content in your inbox daily

Follow these simple steps to get the benefits:

1. Scan the QR code or visit the link below

https://packt.link/free-ebook/978-1-80461-928-5

2. Submit your proof of purchase
3. That's it! We'll send your free PDF and other benefits to your email directly

www.ingramcontent.com/pod-product-compliance
Lightning Source LLC
Chambersburg PA
CBHW080630060326
40690CB00021B/4878